CONFEDERATE CANNON FOUNDRIES

by

LARRY J. DANIEL

RILEY W. GUNTER

PIONEER PRESS, Union City, Tennessee.

Library of Congress Card No. 76-62878.

CONTENTS

PREFACE

The Philadelphia *Inquirer* of March 21, 1862, concurred with General George McClellan's assertion that the Civil War would be "an artillery war." "General McClellan's army is provided with an unusually large number of field batteries of the best and most approved kind . . . . To meet this superb array of rifled guns, what have the Rebels? . . ," continued the *Inquirer*. "This greater preponderence of artillery must gain the day. The Richmond *Examiner* may eulogize shotguns, and Jeff Davis may call out citizens between ten and a hundred if he likes; but, without the magic of artillery, numbers will but add to the disorder and make the panic more fatal." To sustain a protracted war the South would, indeed, need cannon—lots of cannon.

To what sources could the newly born Confederacy turn for a supply of artillery? There was the pre-existant supply, that is, cannon already in the hands of the states and militia companies and those captured at the surrender of the various U.S. installations. The evacuated Gosport Navy Yard at Norfolk, Virginia, and the U.S. forts and arsenals yielded the Confederates in the neighborhood of 1750 cannon, mostly heavy seacoast artillery.[1] In addition, the states, notably Virginia, the military academies, and the few organized militia units, possessed about 400 field pieces.[2] This initial supply of readily available cannon allowed the Confederacy valuable time to gear its foundries to war production. Yet, a large percentage of these guns, especially the field pieces, were so old and obsolete as to be nearly useless and would soon need replacing. Also, hundreds of the heavy guns captured were of inadequate caliber, smoothbore 24, 32, and 42-pounders. Old smoothbore guns, even when converted to rifles, were weak and unreliable, especially against ironclad vessels.[3] Not only would old guns have to be rifled, more powerful cannon would have to be manufactured, also.

Guns captured from the Federal Army also provided a lucrative, though undependable source. Yet, it would be the summer and fall of 1862 before the artillery prizes of the Peninsula Campaign, Second Manassas, Bragg's Kentucky-Tennessee Campaign, and Harpers Ferry, totaling in excess of 250 pieces, would be realized.[4] Nor were the blockade runners of much more help. Only 129 guns entered the Confederacy by this route through January 1863. This supply was to decrease sharply as the Federal Navy tightened its grip on the Southern coast.[5]

The Richmond Government, in desperate need of an immediate supply of reliable artillery, turned to its own private citizens for help. Two of the nation's seven foundries casting cannon for the United States just prior to the war were located in the South.[6] These were the Tredegar Iron Works of Richmond and the smaller Bellona Foundry, located on the James River thirteen miles upriver from Richmond. The Tredegar works, the Krupp of the South, would prove to be the backbone of Confederate ordnance production. In 1966 Charles B. Dew published his study on that facility entitled *Ironmaker to the Confederacy: Joseph R. Anderson and the Tredegar Iron Works* (New Haven and London: Yale University Press, 1966). Dew's book, definitive though it was on Tredegar, was not the final word but rather the beginning of a relatively untouched facet of Civil War history.

There were many foundries scattered throughout the South that could, and did, convert from peacetime to war production. Such firms as Quinby & Robinson of Memphis, T. M. Brennan & Company of Nashville, and Noble Brothers & Company of Rome, Georgia, novices at cannon making, tooled up for production. The Confederate Government, destined to play an ever increasing role in the Southern war machine, eventually developed its own foundries, four for field guns and one for heavy cannon. Emory Thomas' statement that "Confederate manufacturers accomplished little short of an industrial revolution" can nowhere be more vividly displayed than in the field of cannon production.[7] During the war Southern foundries produced over 2300 cannon—half cast by Tredegar and half by the improvised and Government foundries. This study relates the stories of foundries, hitherto unheralded with the achievement and failures of their principals.

For the sake of organization the book is divided into an Introduction and four parts. Part I concerns the established foundries, those facilities involved in cannon production prior to the war. Part II consists of the improvised foundries, existing private plants converted to meet an immediate wartime need. Part III deals with the Government foundries. Part IV, the Appendix, contains a Table of Cannon Cast at J. R. Anderson & Company (Tredegar Foundry). Some foundries either produced so few cannon, or records of their activities are so scant, that they are relegated to the Appendix in A Directory of Confederate Cannon Manufacturers.

Examples of surviving Confederate cannon have been mentioned throughout the book, although no effort has been made to catalogue and describe all surviving weapons. A book on Civil War field artillery, the work of which spans a number of years and the combined talents of three notables in the field—James Hazlett, Hume Parks, and Edwin Olmstead—is soon to be published. For more information on Civil War artillery and specific facts on surviving guns, the authors refer the reader to this work. Another important reference in this respect is

Warren Ripley's pioneer work *Artillery and Ammunition Of The Civil War.*

One point of information. Surviving cannon owned by the National Park Service are occasionally moved from park to park. The mobility of cannon in private hands is even greater. The reader should assume, therefore, that the locations mentioned for surviving Confederate cannon are those at the time of cataloguing. Unfortunately, examples of many foundries, like the firms themselves, have long since passed from existence.

NOTES

[1] "Report of Ordnance and Ordnance Stores on Hand at the Gosport Navy Yard on the 21st of April, 1861 . . . ," Document No. 25, *Documents of the Virginia House of Delegates 1861–62,* Virginia State Library; *The War of the Rebellion, A Compilation of the Official Records of the Union and Confederate Armies* (Washington: Government Printing Office, 1880–1901), Series 4, I, 227–228. Hereafter cited as *O.R.* All references are in Series 1 unless otherwise noted; Graham Daves, "Artillery at the Southern Arsenals," *Southern Historical Society Papers,* XII (Jan.-Dec. 1884), 360.

[2] *O.R.,* Series 3, I, 903, Series 4, I, 63, 386–388, 740; Allen D. Candler, ed., *The Confederate Records of the State of Georgia* (Atlanta: Charles P. Byrd, 1910), III, 41; Stanley F. Horn, ed., *Tennessee's War 1861–1865* (Nashville: Tennessee Civil War Centennial Commission, 1965), 19; Warren Ripley, *Artillery and Ammunition of the Civil War* N.Y.: (Van Nostrand Reinhold Company, 1970), 22.

[3] *O.R.,* XIV, 865.

[4] *Ibid.,* XI, pt. 2, 498, XII, pt. 2, 558, XVI, pt. 1, 1094, 1097, XIX, pt. 1, 141.

[5] *Ibid.,* Series 4, II, 382.

[6] Other foundries casting cannon included Cyrus Alger of Boston, the Ames Manufacturing Company of Chicopee (Springfield), Massachusetts, Fort Pitt Foundry of Pittsburgh, the West Point Foundry of Cold Springs, New York, and the U.S. Naval Gun Foundry at the Washington, D.C. Navy Yard.

[7] Emory M. Thomas, *The Confederacy as a Revolutionary Experience* (Englewood Cliffs, N.J.: Prentice-Hall, Inc., 1971), 87.

ACKNOWLEDGMENTS

The authors wish to gratefully acknowledge the assistance of the staffs of the National Archives; the Alabama, Georgia, Tennessee, Mississippi, and Louisiana State Archives; the Virginia State Library; the Virginia Historical Society; the Louisiana Historical Society; the New Orleans, Memphis, and Vicksburg Public Libraries; West Point Museum; John M. Browning Memorial Museum; and the Shiloh, Petersburg, and Gettysburg National Military Parks. Also, to *Civil War Times Illustrated,* our thanks for permission to use themes and sources previously used in Larry J. Daniel's "Manufacturing Cannon in the Confederacy," XII (Nov. 1973), 4–10, 40–46. Likewise, our thanks to the *West Tennessee Historical Society Papers* for the use of information in Larry J. Daniel's "The Quinby & Robinson Cannon Foundry at Memphis," XXVII (1973), 18–32.

For information, photographs, and manuscript suggestions individual credits go to the following: Troy L. Church, Winston-Salem, North Carolina; John McCaa, Anniston, Alabama; Bill Wright, Jackson, Mississippi; Robert Dunnovant, Jr., Athens, Alabama; Walter R. Averett, Parkersburg, West Virginia; J. C. O'Neal, Vicksburg, Mississippi; and Blaine Russell, Vicksburg, Mississippi.

For many of the photographs of surviving cannon and their inscriptions the authors are particularly indebted to Tarry Beasley of Memphis. Also, particular expressions of thanks to Warren Ripley of Charleston, South Carolina, Richard B. Cohen of Memphis, Tennessee, Hume Parks of Nashville, Tennessee, and Edwin Olmstead of Mount Holly Springs, Pennsylvania. When any information was needed, these gentlemen could be relied upon to respond promptly. Their numerous letters provided both valuable information and encouragement for completing the following pages.

Memphis, Tennessee

L.J.D.
R.W.G.

INTRODUCTION

The casting of cannon was a long and tedious process involving four basic operations—molding, casting, cooling, and finishing. Molding consisted of placing a pattern of the gun to be cast in a box, termed a flask, which was arranged to allow the escape of air and steam. (See photo page 21. Sand was then rammed around the pattern, which, when removed, would leave a cavity into which the melted iron could be poured. When the complete mold was placed together it was assembled and thoroughly dried.

The mold was then placed in a pit and readied for the casting stage. Cannon were made of either iron or bronze (sometimes termed "brass"). Bronze was a mixture of nine parts copper to one part tin. There were no foundries in the South that could cast steel guns. The appropriate metal was melted in a furnace and then poured into the mold, which was filled from the bottom up. As the mold filled a workman agitated the liquid metal with a long rod so as to bring impurities to the top. Filling the mold took only five to ten minutes. The entire flask was then removed for the cooling process. Cooling, especially in heavy seacoast guns, created stress and strain and in an effort to prevent this as much as possible, the flask was surrounded by sand and charcoal. The flask was left in the pit to cool for three to six days, depending on the size of the casting. When sufficiently cooled, the flask was removed from the pit.

Then began the finishing process. The rough casting was placed in a lathe and the exterior usually turned and finished. The exacting work of boring the interior took several weeks for the tubes of larger cannon. Following boring, the trunnions were machined to size. Before removal the exterior was finished and the vent drilled.[1]

Even under the best of circumstances it took some 400–500 hours of labor to complete a 10-inch columbiad weighing 19,000 pounds. It took the Tredegar Iron Works a minimum of one month to cast, finish, and mount such a weapon. For the larger Brooke guns it took the Selma Naval Ordnance Works in the neighborhood of 1000 hours for completion.

During the 1840's and 50's two revolutionary ordnance changes took place. The first was the development of the Rodman method of casting. The old method of casting a gun solid and drilling out the bore worked well with small cannon, but often developed cracks and weak spots in larger tubes. In the 1840's T. J. Rodman of the U.S. Ordnance Department worked out an agreement with the Fort Pitt Foundry at Pittsburgh to cast guns on a hollow core with water kept flowing in and out. In the old method the weapon was cooled from outside inward. Each successive ring to cool would thus tend to shrink away from that which had cooled, producing voids and tension cracks. The Rodman method of cooling was from the inside out, so that the successive shrinkage of each layer in turn produced compression instead of tension. The result was a much improved technique which enabled stronger and heavier guns to be cast. Yet, Joseph R. Anderson at Tredegar and Junius L. Archer at Bellona, the South's two established cannon foundries, failed to adopt the method. In 1859 the Federal Government refused to accept any further weapons not cast by the Rodman method, thus cancelling out Federal contracts with both firms.[2]

The refusal of Anderson and Archer to adopt this revolutionary technique, due primarily to their strong Northern prejudice, seriously affected the production of heavy ordnance for the Confederacy. The largest seacoast cannon that could be cast in a Southern foundry at the beginning of the war was of 10-inch caliber. The term "inch" was a reference to the bore diameter of a gun. The term "pounder" referred to the weight of the projectile fired. Under the Rodman method of casting, Northern foundries were turning out 15 inch and even a few 20-inch guns during the war. These were precisely the types of cannon needed to stop the heavily armored Union ironclads. The Tredegar and Selma plants eventually turned out a total of eleven 11-inch guns, which was the largest gun actually used by the Confederates. Tredegar had converted to the Rodman method by the end of the war and cast, but never completed, two 12-inch guns under this system. However, the conversion simply came too late to be of any value to the Confederacy.

The other revolutionary ordnance change was the rifling of cannon. In the early 1800's all cannon were smoothbores. The term "smoothbore" referred to the smooth internal surface of the barrel. Between 1839 and 1844 the U.S. Ordnance Department adopted a full range of bronze smoothbore field pieces. The most common weapons in use were the Model 1841 6-pounder gun and 12-pounder howitzer. Guns were long barrel weapons designed to throw solid shot with a heavy charge of powder at an effective range of about 1500 yards. Howitzers were first put in use by the French around 1800. They were smoothbore light weapons with shorter barrels designed to fire an exploding shell rather than a solid shot. Howitzers, which had a higher trajectory than guns, were fired by concentrating a light charge of powder at the rear of the tube by means of a chamber. The effective range of a 12-pounder howitzer was about 1000 yards.

For many years artillerists had envied the range and accuracy of the rifled small arm. The rifled musket could

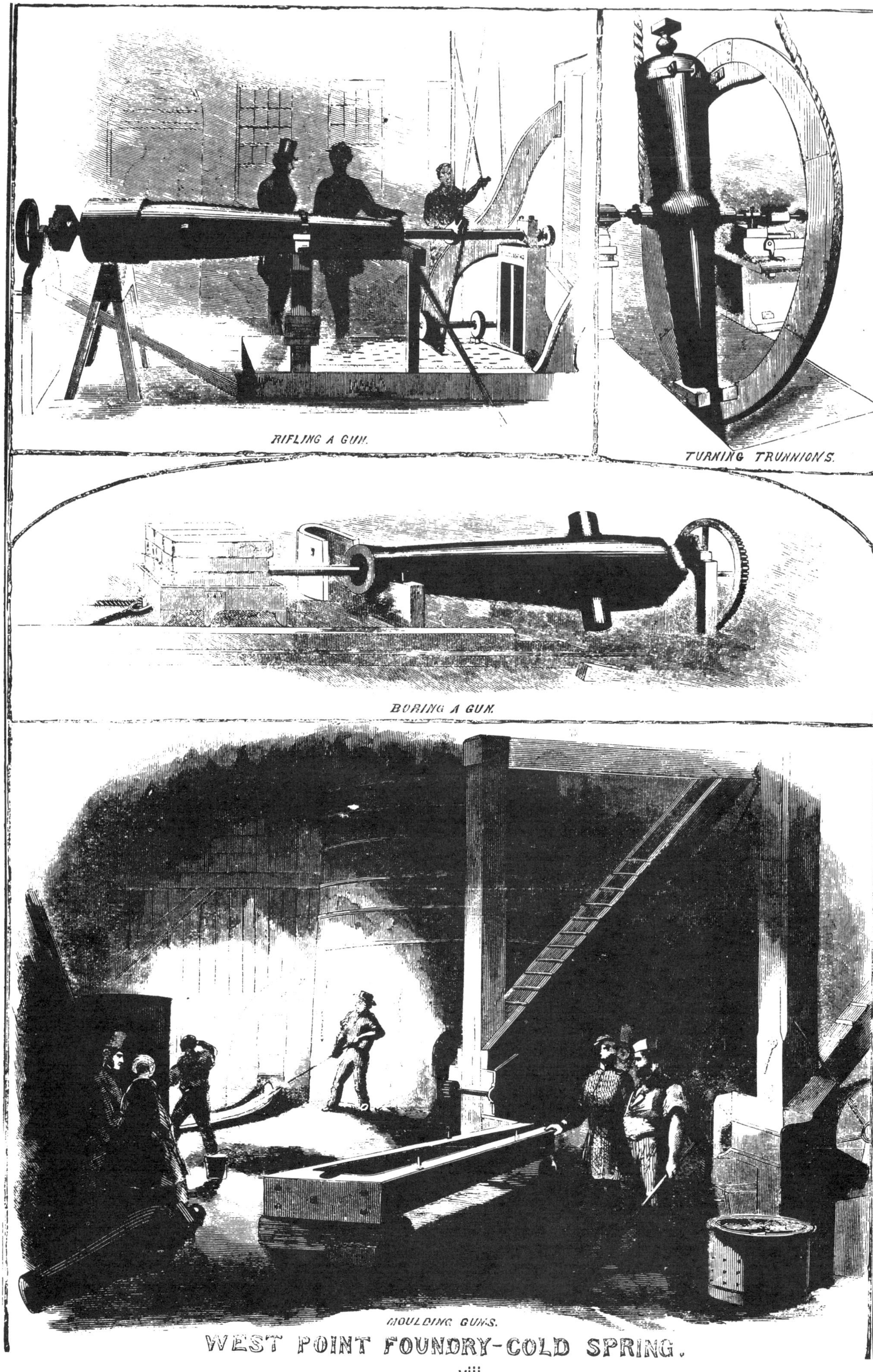

Steps in the Manufacture of cannon at the West Point Foundry, Cold Springs, New York.
(Harpers Weekly, Sept. 14, 1861)

The Fort Pitt Foundry, Pittsburg, Pa.

(Harpers Weekly, Aug. 23, 1862)

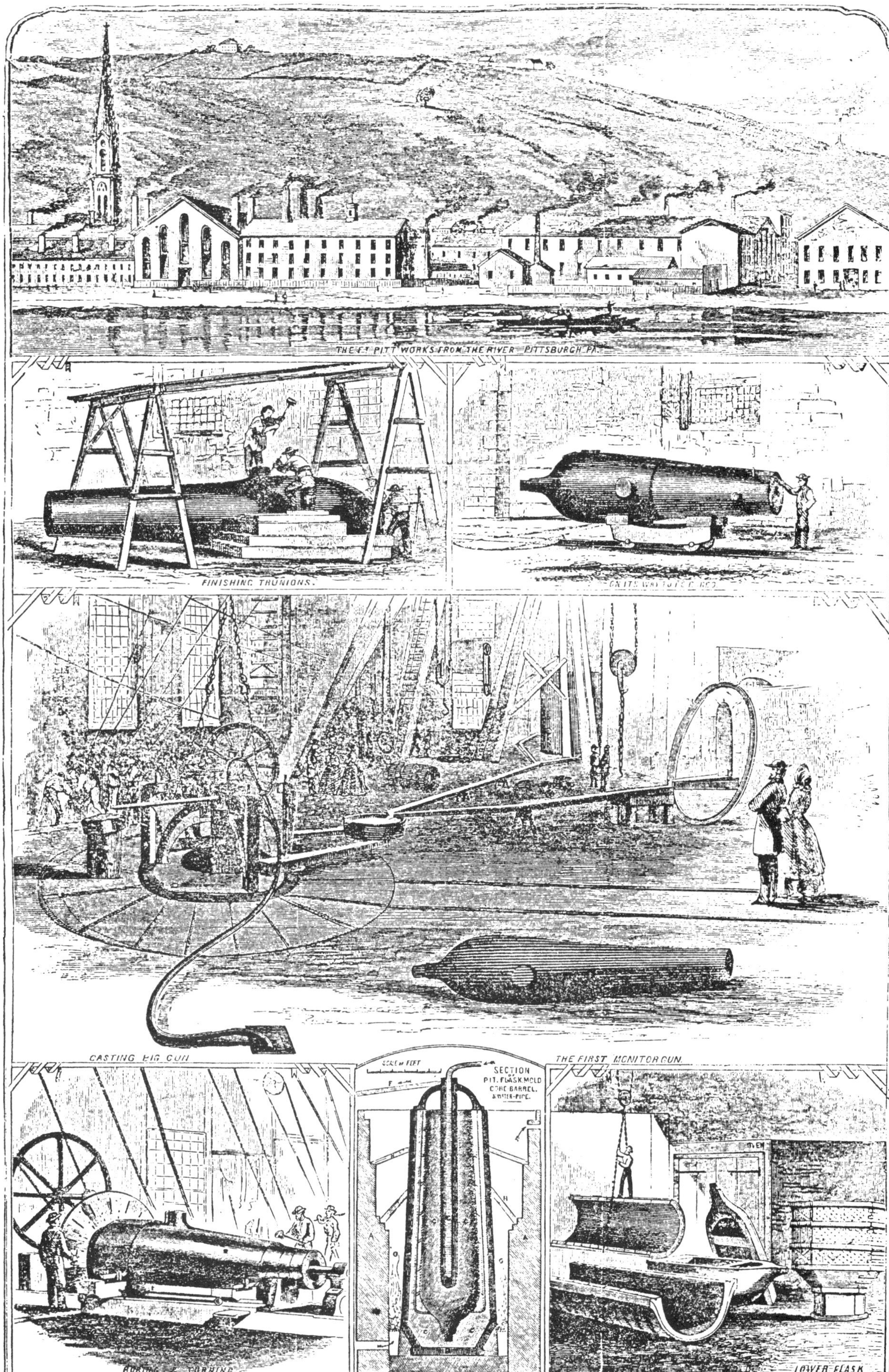

From Harper's Weekly, *August 23, 1862.*

effectively fire 300–400 yards instead of the 50–100 yards of the smoothbore musket. The rifling of cannon proved much more difficult. For years no one was able to design a shell which could be easily loaded and yet fit the bore tightly enough to take the spin which resulted from the grooves of the rifled cannon bore. Finally, in the mid-1850's, a number of designers succeeded almost at the same time. Thus was born a whole new generation of rifled cannon.

Rifled cannon were the wave of the future. They had much greater range, accuracy, and penetrating power than smoothbores. Since rifled guns had longer range than smoothbores they could be deployed further back in the lines, relatively free from capture and harassing fire.

Unfortunately, rifled field pieces had several drawbacks. Rifled shells often buried themselves in the ground which muffled the blast effect. The Confederates were never able to develop a truly effective rifled shell. Canister, the most deadly type of projectile, could not be fired very effectively from rifled guns. The greatest disadvantage, however, was that in the wooded terrain in which most Civil War battles were fought, the added range of rifled guns was of little value. In that situation smoothbores were more effective since they could fire larger projectiles and more canister with deadly effect. Thus, during the Civil War, there was seen value in both smoothbores and rifled cannon.[3]

Despite a great deal of experimenting with rifled cannon during the 1850's, it took time for the technique to catch on. The Federal Government sent some smoothbores back to the foundries to be rifled. Many heavy smoothbores reworked by rifling were also reinforced at the same time; yet the principle as applied to field ordnance was still little known at the turn of the decade. In the field manual of 1860 no official reference was made to the rifling system.

The rifling technique was also applied to heavy seacoast guns. A number of old smoothbore weapons captured in the various U.S. forts and installations were hurriedly sent to Southern foundries in a makeshift attempt to utilize existing material. On occasion rifled smoothbores gave a creditable performance. At best, however, it proved to be only a stopgap method until larger and more powerful guns could be cast. Since production of heavy guns never kept pace in the South, many old converted rifled guns were still in use at the end of the war. Yet, the Confederates attempted to replace them with 8 and 10-inch columbiads and later with Brooke rifles as much as possible.[4]

The columbiad first appeared in 1811. Originally they were chambered like howitzers, but this feature was dropped in later models. The columbiad saw use in the War of 1812 for firing solid shot but was set aside in 1821. The weapon appeared again in 1844 with a larger bore and considerably different appearance. Two models, an 8-inch and 10-inch, were made and were similar other than size. The 8-inch model was over ten feet in length, weighed in excess of 15,000 pounds, and had a range of 2½–3 miles. Machinery for rifling columbiads was completed at the Tredegar plant early in the war. The experiment, by and large, proved to be a failure due to frequent bursting at the breech. General Beauregard later rifled and banded some columbiads in Charleston, and this experiment proved more successful.[5]

In consequence of the frequent use of the term "battery" in this work, the authors feel obliged to make a brief comment. The prescribed make-up of the standard U.S. battery of 1861 consisted of six guns, six caissons, a battery wagon and forge. The four gun battery appeared early in the Confederate field artillery. Although this was not an economical arrangement, the War Department made no objection to it, and, as the army was hard-pressed from the beginning to equip and maintain six gun batteries, few appeared after the spring of 1862. In the Army of Northern Virginia the six gun battery had all but disappeared by the Peninsula Campaign. The majority of batteries in the Confederate Army of the Mississippi had six guns at the battle of Shiloh in April, 1862, but were reduced to four guns shortly after the battle.[6]

The number of forges and battery wagons were likewise reduced. After Gettysburg Lee's entire Artillery Corps had only 8 battery wagons and 32 forges to accompany 62 batteries.[7] In March, 1864, the Army of Tennessee alloted only one forge per battalion of three batteries.[8] After May, 1864, the production of battery wagons was ceased, "quartermaster wagons with good tarpaulin covers being preferred." [9]

NOTES

[1] Ripley, *Artillery and Ammunition of the Civil War,* 243–245; Memphis *Appeal,* April 20, 1861; Richmond *Enquirer,* September 28, 1861; Charles B. Dew, *Ironmaker to the Confederacy: Joseph R. Anderson and the Tredegar Iron Works* (New Haven and London: Yale University Press, 1966), 110–111.

[2] *Ibid.,* 45, 49–50; Ripley, *Artillery and Ammunition of the Civil War,* 78.

[3] Harold L. Peterson, *Round Shot and Rammers* (New York: Bonanza Books, 1969), 92; Jac Weller, "The Field Artillery of the Civil War," *The Military Collector and Historian,* V (September 1953), 65–66; Ed McDonald, Wayne Praeder, and Mike Bianchi Rossi, "Rifle Musket: American Firepower and Tactics 1861–1865," *Conflict,* VII (June 1974), 14–17.

[4] Ripley, *Artillery and Ammunition of the Civil War,* 17–18.

[5] *Ibid.,* 71–73, 83–84, 368.

[6] E. P. Alexander, "Confederate Artillery Service," *Southern Historical Society Papers,* XI (1883), 99; Jennings C. Wise, *The Long Arm of Lee* (New York: Oxford University Press, 1959), 164, 186, 201, 202, 207; Edwin C. Bearss, "Artillery Study Shiloh National Military Park," Project No. 17.

[7] Wise, *The Long Arm of Lee,* 706–707.

[8] *O.R.,* XXXII, pt. 3, 695.

[9] Orders and Circulars, Ordnance Department, Augusta Arsenal, National Archives Record Group 109 (War Department's Collection of Confederate Records).

A Section of the Palmetto (S.C.) Light Artillery Regiment on duty in Charleston in 1863.
(Library of Congress)

THE ESTABLISHED FOUNDRIES

The Tredegar Iron Works, Richmond, Va., photographed in 1865 *(Library of Congress)*

J. R. ANDERSON & COMPANY

The famous Tredegar Iron Works of Richmond, Virginia, evolved from a small forge and rolling mill established in the mid-1830's by Francis B. Deane, Jr. and a group of associates and named after the Tredegar works in Wales. The firm did not operate successfully until 1841 when it came under the leadership of Joseph R. Anderson. Anderson—Virginia-born, West Point graduate, and later ardent secessionist—straightened out the tangled business affairs of the firm and expanded operations to compete in Northern and foreign trade markets. In 1843 the firm voted to dispose of the property to Anderson under a five year lease and in 1848 he assumed complete ownership. The works, although locally always known as Tredegar, soon came under the heading of J. R. Anderson & Company. By 1860 the plant, employing some 800 free and slave laborers, was one of the largest in the nation.[1]

Tredegar had a long heritage of cannon producing prior to the Civil War. Initial difficulties were experienced when in 1843 the U.S. Navy rejected forty of the first contracted 100 guns due to the failure of five cannon to pass firing tests. The "disaster of '43", as the rejection became known, was a severe blow to the plant. These difficulties were eventually overcome by the use of better iron and improved casting methods. Between 1844 and 1860 Tredegar delivered 881 pieces of ordnance to the

A 6-pounder bronze smoothbore cast by Tredegar Feb, 1861, for the State of North Carolina during the tension filled days just prior to the war. It survives today at Shiloh National Military Park.

(Tarry Beasley)

Federal Government, not one of which failed to pass firing tests.[2]

Tredegar financial affairs were in a serious state at the turn of the decade with the loss of Federal ordnance contracts due to the firm's failure to adopt the Rodman method of casting. (See Introduction) However, financial assistance was soon to come in the form of a relatively untapped area—cannon and munitions for Southern states. With war clouds on the horizon and the sale of Northern cannon cut off, Southern states began taking steps for their own defense. Many states made inquiries to the Tredegar plant and some placed sizeable orders for cannon. During the tension filled days just prior to the Civil War Tredegar delivered fifty-six cannon (fourteen field pieces, seven seacoast mortars, thirty-five heavy guns) under contract to South Carolina, Georgia, Alabama, and Mississippi. Some of these weapons, those destined for South Carolina, were soon to be turned against Fort Sumter in Charleston Harbor.[3]

Anderson was elated over the surrender of Fort Sumter on April 14, 1861. The capitulation was marked in Richmond by a mass parade and gun salute, all organized by Anderson to encourage Virginia delegates toward secession. Anderson soon offered the plant to the newly organized Confederate Government by lease or purchase. The authorities declined, however, and Tredegar remained in private hands throughout the war.[4]

Virginia soon imposed an embargo on all state orders placed with Tredegar. Guns awaiting shipment on railroad cars were detained. Virginia rushed a number of old guns to the plant from her Richmond Armory to be reamed out to a good bore and rifled. The War Department finally clamped down on these state orders in the summer and fall of 1861 and thereafter the firm cast solely for the Confederate Government.[5] Tredegar did cast a few field guns in time to participate in the opening conflict at Manassas.[6]

Faster production was needed. Shortly after First Manassas General Joseph E. Johnston wrote President Davis urging an increase in the field artillery "by borrowing from the States or by casting, especially at Richmond." [7] In August Colonel William N. Pendleton, Johnston's artillery chief, was sent to Richmond to expedite the casting of guns in that sector. "Have had a great deal to do here in pushing up artillery preparations," he reported. "Will have several batteries at Manassas next week, and several others the week after, etc. Several difficulties obstruct the way. Want of tin to make brass; only one good foundry here, deficiency of hands, etc . . . . Still, by hook or crook we get along. "[8]

As the war continued, the need for heavy guns also increased. "The winter of 1861–2 was the darkest period of my department," recalled Chief of Ordnance Josiah Gorgas. "Heavy guns were called for in all directions—the largest guns for the smallest places." [9] Tredegar concentrated on the casting of heavy guns throughout the fall and winter of 1861. Thirty-nine guns were shipped by rail to all parts of the Confederacy between October and December.[10] The Richmond firm soon had a casting capacity of six to eight field guns and six heavy guns per week.[11]

As early as July Anderson informed the Chief of Ordnance that machinery had been developed for rifling columbiads. In August the Army placed a large order for these weapons.[12] "The machinery for boring and rifling cannon . . . are kept constantly employed," reported a Richmond correspondent. "During our visit on Wednesday, a monster 10-inch columbiad was being rifled—a process which would require about two days." [13] Since the full bore 8 and 10-inch guns could not stand the strain of rifled projectiles, Tredegar bored the 8-inch pattern to 5.82 inches, the caliber of a 32-pounder, and the 10-inch to 6.40 inches, the caliber of a 42-pounder.[14]

Anderson was less enthusiastic over a Government order for 15-inch columbiads, but since his plant was the only one capable of undertaking such a project, the arrangements were begun.[15] The 15-inch gun was developed at the Fort Pitt Foundry under the Rodman method of casting.

Tredegar's most successful heavy ordnance innovation was the Brooke rifle. The gun was developed by Lieutenant John Mercer Brooke, formerly of the United States Navy, and later head of the Confederate Ordnance and Hydrography Department. Brooke was the Navy's counterpart to Josiah Gorgas. The Brooke, like the Northern-made Parrott rifle, was a cast iron gun with wrought-iron reinforcing. The Brooke was much heavier, however, had a rifling similar to that of the English Blakely, and could be single, double, or triple banded. A double-banded 7-inch Brooke rifle, weighing over 15,000 pounds, could hurl a projectile over 4½ miles.[16] Fourteen Brooke were turned out between September, 1861, and March, 1862.[17]

Tredegar's large production of heavy ordnance required huge stores of reliable iron. Anderson had found only one type of iron suitable for heavy guns, that of the Cloverdale furnace in Botetourt County. To supplement the supply of Cloverdale iron Anderson also contracted with other furnaces, among them Graham, Columbia and Fort. Graham iron, which was reputedly of high quality, proved to be completely unreliable. By November, 1861, three heavy guns cast of Graham metal had burst, one during the heat of battle at Port Royal, South Carolina. The army ordnance inspector promptly rejected two heavy guns cast of Graham iron.[18]

Difficulties due to poor iron and shoddy workmanship continued to plague Tredegar ordnance production. Four or five Tredegar heavy guns, including some rifled columbiads, burst during the month-long siege of Yorktown, thus creating shouts of distrust from many quarters.[19]

Joseph R. Anderson *(Library of Congress)*

A Tredegar Brooke rifle photographed at Dutch Gap Canal in April, 1865. Tredegar cast approximately 112 Brookes of all types, about 11 of which survive. (*Library of Congress*)

"There must be something very rotten in the Ordnance Department," blurted General D. H. Hill in April, 1862. "It is a Yankee concern throughout, and I have long been afraid that there was foul play there. . . . The metal of which the new guns are made is of the most flimsy and brittle character, and the casting is very bad." [20] Gorgas denied disloyalty at the plant, but frankly conceded that ". . . that establishment has not sustained its old reputation." [21]

Deliveries of field guns and battery equipment increased in the fall of 1861. By September the firm had begun monthly deliveries of ten to fifteen field guns. Between September and December Tredegar made deliveries of twelve 6-pounder rifles, sixteen 6-pounder smoothbores, and twenty-four 12-pounder howitzers, a total of fifty-two pieces. Tredegar carpenters, formerly freight car builders, turned out seven traveling forges and dozens of gun carriages between August and December.[22]

The severe copper shortage halted the production of all bronze ordnance in May, 1861. Until copper could be

TREDEGAR COLUMBIAD

The 8-inch columbiad above was cast by Tredegar on October 13, 1862. It was shipped to Mobile, Alabama, where it was dismounted in 1865 by concentrated Federal artillery fire. It survives today at New Orleans.
(La. Historical Society)

furnished Anderson & Company was forced to cast iron ordnance. In August the Ordnance Department placed orders for 106 iron field pieces—12-pounder howitzers, 3-inch rifles, and 6-pounder rifles. Bronze ordnance production was resumed in December but a number of iron weapons continued to be cast throughout the winter and spring of 1862.[23]

Iron 6-pounders and 12-pounder howitzers differed materially in both weight and shape from the standard bronze weapons of the same caliber.[24] A number of batteries, such as the Rockbridge (Virginia) Artillery and Carpenter's (Virginia) Battery, were armed with Tredegar iron guns prior to the Peninsula Campaign and early reports seemed favorable.[25] Later, however, poor iron and faulty casting techniques created difficulties.

The bursting of an iron Tredegar gun during routine firing tests in March, 1862, caused Gorgas to investigate the problem. He was shocked to discover that iron for a number of 3-inch rifles, iron 6-pounders and howitzers had been melted in a cupola furnace instead of Tredegar's specially constructed gun iron furnace. The War Department severely reprimanded Anderson for this action and warned that all Tredegar field pieces then in service would undergo strict inspections. As a result of these tests a number of iron guns were recast the next month.[26]

The Richmond Government hurried war preparations

Josiah Gorgas, Confederate Chief of Ordnance. *(National Archives)*

TREDEGAR 12-POUNDER HOWITZERS

Top: An 1862 bronze 12-pounder howitzer located at Shiloh.

Below: An 1862 iron 12-pounder howitzer in private hands. *(Tarry Beasley)*

as McClellan's Army of the Potomac inched closer to the capital city. Tredegar made a concerted effort to cast field guns during April, May and June, 1862. The foundry reached its peak monthly capacity in April with the casting of seventy-two pieces.[27] About a month was required before a gun was readied for the field. Some seventy-four completed cannon were delivered in May and June including seventeen 3-inch rifles, twenty-five 12-pounder howitzers (eight iron and seventeen bronze), twenty-three 6-pounder smoothbores (thirteen iron and ten bronze), three 2.25-inch mountain rifles, and six 12-pounder mountain howitzers.[28] The mountain rifles and mountain howitzers were for use in the rugged areas of western Virginia and East Tennessee.[29]

Bell metal was used in the casting of some of the bronze guns.[30] "We can hear whistling of locomotives in and around Richmond, but the sound of a bell has not yet reached our ears," reported a Northern correspondent outside Richmond. "Wonder if they really have cast them all into cannon?" [31] Some old bronze cannon were also melted down to be recast.[32]

During the Peninsula Campaign the Confederate Army fielded seventy-seven batteries with approximately 300 guns.[33] Although specific records are lacking, it would not seem an overestimate to assume that fully one-half of these guns were Southern-made, mostly at Tredegar. It was this supply of guns, furnished at a time before large captures from the enemy or the acquisition of blockade shipments, which in large part accounted for the Confederate victory in that Campaign.

Although Tredegar channeled most of its energies to the casting of standard military cannon, some unique innovations were also turned out. In the fall of 1861 R. S. Williams supervised the construction at Tredegar of a rapid fire breechloading gun invented by himself. The weapon fired eighteen to twenty shots per minute. The first Williams gun saw service at Seven Pines on May 31, 1862. The results were so satisfactory that the Confederate Government had Tredegar complete a battery of these weapons. This particular battery served in Sterling Price's Army of the West. During 1862–63 Tredegar turned out twenty Williams breechloaders. Excessive firing was found to expand the breech which caused a failure in the relocking mechanism, however, and the weapons were later discarded.[34]

One of the primary problems that faced the Confederate Ordnance Department at the commencement of the war was the rapid production of an adequate supply of field guns. This problem was largely solved by the summer of 1862 by captures from the enemy and guns cast by Southern foundries.[35] On June 8, 1862, General Robert E. Lee wrote the Secretary of War discouraging the organization of additional batteries in Richmond and recommended that the horses and men be transferred to other branches of the service. General P. G. T. Beauregard had by that time sent a number of guns to the rear.[36]

No sooner had the difficulty of quantity been solved, however, than the quality of the guns manufactured became of utmost concern. Of the approximately 400 field pieces cast by nine major Southern foundries between September, 1861, and June, 1862, nearly seventy-five percent were of the old Model 1841 smoothbore 6-pounder and 12-pounder howitzer type. These standard weapons had proven themselves in the past and the authorities seemed reluctant to advance to more modern cannon. "Do not fail to urge the making of 12-pounder howitzers," wrote General Joseph E. Johnston after First Manassas. "I have faith in them." [37] At the battle of Shiloh in April, 1862, about eighty percent of the Confederate artillery consisted of 6-pounders and 12-pounder howitzers. In comparison, the Union artillery consisted of nearly fifty percent rifled cannon.[38]

By 1862 important changes were taking place in the

field of ordnance. The 6-pounder and 12-pounder howitzer, weapons which served the Confederates well in the early months of the war, were rapidly being outclassed by the more powerful and long range guns of the enemy. More advanced weapons, such as the 12-pounder Napoleon and Parrott rifle, were being produced in Northern foundries in ever increasing numbers. During the Peninsula Campaign fully one-third of the Federal artillery consisted of 12-pounder Napoleons.[39] By the end of 1862 nearly 600 Napoleons and 370 field Parrotts had been turned out in Northern foundries.[40]

The 12-pounder Napoleon was developed by the French Army in the 1850's under the commission of Emperor Louis Napoleon and named after him. Up to that time field pieces consisted of either guns or howitzers. Guns were heavy weapons designed to throw solid shot at great distances with a heavy charge of powder. Howitzers were lighter weapons designed to fire shells at shorter distances with a reduced charge of powder.

The 12-pounder Napoleon gun, or "Napoleon" as it was called, was developed as a hybrid weapon. It was designed to fire shells or canister as the howitzer and yet retain the range of the gun. It was thus termed "gun howitzer" by the French, although technically the weapon was a gun. Its advantages were many. Its range and accuracy were greater than the 6-pounder and 12-pounder howitzer. Its weight, 1227 pounds, was 500 pounds lighter than the old Model 1841 12-pounder. Also, when firing canister at 300 yards, it had a devastating effect upon advancing infantry.[41] As Gorgas later recalled: "At the distance which the serious work of the artillery was done, it was an overmatch for rifled artillery."[42]

TREDEGAR NAPOLEONS

Top: 1863 bronze Tredegar 12-pounder Napoleon at Gettysburg.

Below: An iron 12-pounder Napoleon. Note the resemblance to Parrott rifles. Only four of these weapons are known to survive. *(Tarry Beasley)*

The Parrott rifle was developed by Robert P. Parrott of the West Point Foundry in 1859–60. The gun was a cast iron rifle with a reinforcing jacket shrunk around the breech. The gun was made in two field models—a 10-pounder and 20-pounder. The 10-pounder weighed 900 pounds and had a bore diameter of 2.9 inches, although later Union models were enlarged to 3.0 inches to permit interchange of ammunition with the 3-inch rifle. The 10-pounder Parrott had a range of 3200 yards, twice that of Napoleons, and was particularly effective for long range use.[43]

Confederate impassiveness to these ordnance changes was not due totally to a lack of vision. Prior to the war, for example, an Ordnance Commission from the State of Virginia had seen the advantages offered by the new Parrott rifle. The State purchased thirteen 10-pounder Parrotts, some of which were used with great effectiveness at the battle of Big Bethel.[44] Beginning in the late fall of 1861 and throughout the summer of 1862 Tredegar cast some 10-pounder Parrotts.[45] Yet, early difficulties encountered by the Confederates in perfecting a rifled shell discouraged the large scale use of rifled guns.[46] Delay in converting to more modern ordnance, however, was largely due to the conservative nature of Confederate military leaders and the Ordnance Department. When efforts were finally made to convert to more modern weapons, it was only in response to a more powerful Federal armament. In short, the authorities did not act—they reacted.

In July, 1862, the Ordnance Department ordered Tredegar to cease the casting of 6-pounders and 12-pounder howitzers and prepare the patterns for the 12-pounder Napoleon. Lieutenant Colonel R. S. Andrews obtained possession of a Federal book on field artillery and republished it in a volume of his own. From these plans Tredegar was able to make the weapon.[47] The first pieces were transitional but the weapon was soon perfected to satisfaction.[48] There were varying reports as to the performance of the gun. Some officers praised the Tredegar weapon, stating that it jarred less than its Federal counterpart.[49] Yet, other officers candidly admitted that the Northern Napoleon was superior.[50]

On July 26, Tredegar cast its first 30-pounder Parrott which was patterned after a Federal gun captured by the Confederates at First Manassas and dubbed "Long

A Tredegar 20-pounder Parrott rifle cast in April, 1864, and located in Marietta, Ohio. *(Walter R. Averatt)*

Tom." [51] Delivery of the first two 30-pounders did not come until September.[52]

Captured Federal ordnance provided a lucrative source for the Army of Northern Virginia. Nine 10-pounder Parrotts were taken at First Manassas.[53] The Peninsula Campaign netted an additional eighteen 10-pounder Parrotts as well as seventeen Napoleons.[54] The victory at Second Manassas yielded additional weapons.[55] For the Maryland invasion in September, 1862, Lee had available six 20-pounder Parrotts, forty-five 10-pounder Parrots, and twenty-eight 12-pounder Napoleons.[56] Yet, the field artillery still consisted chiefly of 6-pounders and 12-pounder howitzers, a factor which considerably lessened the fighting ability of the Army. Confederates were greatly outmatched at Antietam in their artillery, with Federal 20-pounder Parrotts dominating the field. General S. D. Lee later recalled that "Sharpsburg was Artillery Hell." [57]

Insufficient raw materials slowed production of both Napoleons and Parrotts during the last half of 1862. Each Tredegar Napoleon required over 1000 pounds of copper, a commodity in constant short supply. Transportation of copper from the Ducktown mines near Chattanooga involved the use of at least three railroads and the severence of this vital link was a matter of constant concern.[58] Iron was also in short supply as poor weather conditions hampered operations at the Cloverdale furnace throughout the winter of '62–'63.[59]

The modernizing of the field artillery became a chief concern of Lee in late 1862. "During the past campaign I have felt, in every battle, the advantages that the enemy possessed over us in the artillery," Lee informed the Secretary of War. "This arose in part from their possessing more experienced artillerists and better prepared ammunition, but consisted chiefly in better guns." Lee urged the Ordnance Department to improve the situation, concluding that "the contest between our 6-pounder smoothbores and the 12-pounder Napoleons of the enemy is very unequal. . . ." [60] Gorgas, in anticipation of Lee's request, had in November ordered future production of all field guns limited to Napoleons and 10 and 20-pounder Parrotts.[61]

Tredegar delivered a few 20 and 30-pounder Parrotts to Lee before the end of 1862. Four 20-pounders were shipped in October and two 30-pounders in late November.[62] The 30-pounders were hurriedly prepared in emplacements on Marye's Heights in time to see action at the battle of Fredericksburg in December. The guns were used with effectiveness for several hours during the battle, but the rapid rate of fire overheated the weapons and both burst, one on the thirty-ninth round and the other on the fifty-fourth round. Generals Lee and Longstreet were standing near one gun when it burst and barely escaped injury.[63] "Happily it did not send fragments flying about, and no one was hurt," reported a staff officer. "The Immense breach [sic] just appeared to have split into a dozen pieces of various sizes and then fallen heavily to the ground. We were rather glad to have done with such a piece of metal." [64]

Confederates continued to favor Northern-made Parrotts over Tredegar Parrotts throughout the war. In April, 1863, a Board was ordered to evaluate the performance of Tredegar 20-pounder Parrotts to ascertain "whether those captured are 'more esteemed' than the 10 and 30-pounder Parrotts." [65]

Copper shortages plagued production capabilities at Tredegar during January and February, 1863. In March an anxious Lee informed President Davis that to replace the 6-pounders in his Army with 12-pounder Napoleons would require seventy Napoleons in addition to the ones then on hand. To augment the Government supply of copper Lee recommended that the old 6-pounders be sent in, melted down, and recast into light 12-pounders.[66] "We are pushing ahead the Napoleons for General Lee and have melted nearly the whole of the old guns," wrote Anderson. "We will require at once either copper or more old guns." [67]

The flow of Tredegar Napoleons to the field was a slow process. Nearly all of the 6-pounders of the 2nd Corps had been sent in by late January. Deliveries of guns from the 1st Corps and the Artillery Reserve began in February.[68] By the end of the month the 1st Corps had ninety-six remaining field pieces, including twenty 3-inch rifles (four Tredegar made), eighteen Parrott rifles (all Federal guns), and twenty-one Napoleons (only one of which was a "Richmond Napoleon"). Of the thirty remaining 6-pounders and 12-pounder howitzers sixteen were iron pieces.[69] Since all the bronze 6-pounders were absent, some batteries were down to only one or two guns. Lee, therefore, cautioned that the "reduction in the number of guns cannot be great." [70] By the end of March only eleven 6-pounders remained in Lee's Army, all of which were promptly turned in the next month.[71]

With adequate copper in the form of obsolete cannon, Tredegar production of bronze Napoleons sharply increased during the first six months of 1863.[72] A number of light 12-pounders were delivered to Lee prior to Chancellorsville. By the end of May a fresh consignment of fifteen Napoleons had also arrived. These guns, coupled with the ten Federal Napoleons and rifles captured at Chancellorsville, aided considerably in the artillery modernization.[73] A total of forty-nine Tredegar Napoleons were delivered in time to participate in the spring and summer campaigns, which enabled the replacement of all of Lee's 6-pounders prior to Gettysburg.[74] Iron shortages caused Parrott production to drag, however. Only twelve field Parrott rifles were turned out between January and August, 1863.[75]

Disaster struck the plant in the early hours of May 15, 1863. A fire which originated in the Crenshaw Woolen Mills soon spread to the Tredegar works igniting much of the plant. Before the fire department and bucket brigades were able to bring the flames under control a large portion of the works had been destroyed, including two foundries, one of which had cast cannon since 1842, the gun carriage works, blacksmith shop, pattern shop, and a large section of the machine shop. Yet, many of the vital departments, including the new gun foundry where bronze guns were cast, were saved. It required many weeks before normal operations were resumed in all shops.[76]

The May fire hampered Tredegar ordnance production at a crucial time, just six weeks prior to Lee's Pennsylvania invasion. It must be remembered, however, that it took at least four weeks to prepare a gun for the field. Thus, in terms of the Pennsylvania invasion, it represented only about a two week production loss. This fact must be remembered in balancing Charles Dew's statement that "the Tredegar fire and the defeat of the Army of Northern Virginia [at Gettysburg] were not totally unrelated events." [77] It can further be argued that the absence of Tredegar Parrotts at Gettysburg did not effect the artillery bombardment on July 3rd nearly so much as

A Tredegar 1863 10-pounder Parrott at Shiloh
(Tarry Beasley)

the tactical misplacement of guns, overshooting the target, and the failure to utilize all available weapons.[78]

The critical military situation in Mississippi became an ever increasing concern for the Ordnance Department in the spring of 1863. Some Tredegar columbiads and Brookes were dispatched to Vicksburg and Port Hudson and in April these citadels were issued a belated top priority in receiving Tredegar heavy guns.[79] Yet, the difficulty of shipping heavy cannon so far inland remained a severe obstacle and both positions remained relatively undergunned.[80] Even more serious was the fact that both Tredegar Brookes shipped to Vicksburg burst during the siege.[81] This was all too reminiscent of difficulties experienced in the spring of 1862.

A shortage of iron and the May fire had severely limited Tredegar production of heavy guns. During the reconstruction interim, guns were rifled and banded at the nearby Richmond Naval Ordnance Works.[82] John Brooke of the Ordnance and Hydrography Department remained optimistic concerning Tredegar production potential. He forecast that with the completion of new Tredegar facilities "the number of guns provided in an equal interval will be double what it was before the fire." [83] Casting of cannon was resumed on May 28th but actual production failed to meet Brooke's projections.

Anderson & Company turned out seventy-eight heavy guns during the last six months of 1863, the vast majority of which were 8 and 10-inch columbiads and 7-inch

15-inch RODMAN

Tredegar's pre-war refusal to adopt the Rodman method of casting blocked the production of these monster weapons at the foundry throughout the war. By the end of the war Northern foundries were turning out 20-inch guns. *(Library of Congress)*

A TREDEGAR BROOKE

This triple-banded 7-inch Tredegar Brooke rifle was mounted on Sullivan's Island in Charleston Harbor in August, 1863. It was used in a battle against Federal ironclads on September 3rd. On December 6th the breech developed a crack while firing at a Union monitor. It survives today in Charleston.

(Library of Congress)

Brooke rifles.[84] Smoothbore heavy guns of lighter caliber, 24, 32, and 42-pounders, had proven ineffective against powerful Union ironclads. For effective work of this nature columbiads and rifled guns, especially Brookes, were required.[85] Many of these guns were shipped to the defenses at Charleston, Wilmington, and Mobile.[86] Tredegar Brooke rifles played a decisive role in repulsing the Federal ironclad fleet at Charleston on April 7, 1863.[87]

Disturbing reports of faulty ordnance continued to be received by authorities. A Tredegar 6.40-inch Brooke burst at Charleston in July, 1863.[88] A 7-inch Brooke received at Savannah that same month had to be condemned.[89] In September rebanding and repair work had to be performed on several Charleston Brookes.[90] Difficulties continued with the bursting of a triple banded Brooke on Sullivan's Island in September, followed by the bursting of another Tredegar Brooke in January, 1864.[91]

Tredegar cast its first 11-inch Brooke in December,

1863 and another in February, 1864. It was July before these cannon were completed and dispatched to Charleston. In November a 12-inch gun was cast under the Rodman method. This cannon, and a similar weapon cast in February, 1865, were never completed. The plans for 15-inch columbiads, presented in the opening weeks of the war, never materialized. Anderson's pre-war refusal to adopt the Rodman method of casting blocked the casting of these weapons throughout the war. When the firm finally did convert, it was too late to be of any value to the then dying Confederacy.[92] Northern foundries not only turned out a number of 15-inch guns, but, before the end of the war, also cast some huge 20-inch guns.[93]

Tredegar production of field pieces centered on bronze Napoleons and iron Parrotts during the last six months of 1863. The fire the previous spring and subsequent iron shortages had restrained Parrott production. These difficulties were largely overcome by the fall and the casting of Parrotts sharply increased. Between September and November the plant turned out twenty 10-pounders, fifteen 20-pounders, and four 30-pounders.[94]

Most Tredegar field guns served in Lee's Army of Northern Virginia, but some were also dispatched to the Western Theater. An 1863 listing of the guns in Walker's Division, Army of Mississippi, revealed four Tredegar cannon in use—one iron and one bronze 12-pounder howitzer and two 20-pounder Parrotts.[95] A number of Tredegar guns were recaptured by the Confederates at Chickamauga in September, 1863, including three 12-pounder mountain howitzers, two iron 12-pounder howitzers, and one 3-inch rifle.[96] Eight Tredegar weapons passed through the Atlanta Arsenal during 1864, namely, four mountain rifles, one iron and one bronze 12-pounder howitzer, and two 20-pounder Parrotts.[97]

In November, 1863, the Ducktown copper mines outside Chattanooga fell to Union troops. This brought the production of bronze Napoleons, the favorite gun of the field artillery, to a halt. Fortunately, the blow did not prove crippling. In January, 1864, Tredegar foundrymen developed an iron Napoleon banded with a two-inch thick reinforcing at the breech for extra strength.[98] Although somewhat heavier than the bronze weapon, the gun proved entirely satisfactory.[99] Casting difficulties were experienced in the initial stages but these were soon overcome.[100] The weapons passed firing tests in March and the Ordnance Department was soon giving the production of iron Napoleons preference over all other cannon.[101] Some 120 of these guns were cast before the end of the war.[102]

The iron Napoleon received high praise from artillerymen. A member of the Richmond Howitzers noted after the war that "the iron gun was not only equally safe from explosion, but soon accomplished every purpose against the foe possible with the brass gun and did not create the sharp, piercing ring so severe as not infrequently caused blood to break from the ear of the cannoneer." [103]

Siege operations around Petersburg became a primary concern during 1864. Heavy siege cannon—30-pounder Parrotts, columbiads, and mortars—were rushed to the city. Some light coehorn mortars were also delivered. The plant could manufacture these at the rate of three per week.[104]

The Confederates lost a number of guns through capture during the closing months of the war, including many Tredegar weapons. During Sheridan's Shenandoah Campaign 101 guns were captured, twenty-one of which were Confederate-made. These included sixteen 12-pounder Napoleons (seven bronze and nine iron) and five 3-inch rifles.[105] Probably most, if not all, of these guns were cast at Tredegar. The Confederates also abandoned twenty field guns in the Petersburg trenches, sixteen of which were stamped with the "J.R.A. & CO." (Joseph R. Anderson & Company) marking. These included one 12-pounder howitzer (iron), two 10-pounder Parrotts, and thirteen 12-pounder Napoleons (six bronze and seven iron).[106]

The shortage of skilled labor increasingly affected cannon production. The 1000 man labor force employed in 1861 slowly dwindled to between 400–500 by late 1864. White Tredegar mechanics were organized into a defense battalion which was, late in the war, often called upon to serve in the Richmond defenses. So valuable were the lives of three ironworkers, that it was believed that cannon manufacture would cease if they were killed. During extended periods in the trenches Tredegar workmen often deserted in alarming numbers. "Workmen will not fight and work both," argued Gorgas. "This must be accepted as settled in their minds." [107]

Tredegar turned increasingly to slave labor for its work force. By November, 1864, the free labor force had been cut by more than fifty percent, while the slave population had more than doubled. Ed Taylor, a slave belonging to Anderson, was highly skilled in hammering out the iron bands for Parrott and Brooke guns.[108] In general, however, slaves were never able to fill the vacancies left by the departure of skilled white mechanics.[109]

By 1865 a severe shortage of the right kind of iron had jeopardized both the quantity and quality of Tredegar ordnance. The capture of the Cloverdale furnace by U.S. troops necessitated the use of inferior Liberty iron. Production at the gun mill came to a complete halt by March 1, 1865.[110]

Richmond was surrendered the next month. Federal officers found Tredegar to be in a totally depleted state. The plant lacked any of the finer machinery which had been previously purchased in the North. Some new machines had been made at the works but these were not of the finest workmanship. The gun pits were filled with water and the furnaces were in need of new fire brick.

Captured Confederate cannon lined up for transport after the surrender of Richmond. *(Library of Congress)*

The machine shop, where turning and boring of cannon was done, contained twenty lathes, many of which were of an old pattern. The carpenter shop and some other buildings destroyed in the May, 1863, fire had never been rebuilt. Approximately twenty-five cannon were in various states of manufacture, including two 12-inch Rodman smoothbores, nine 12-pounder Napoleons (iron), three 10-pounder Parrotts, three mortars, three 24-pounder howitzers (iron), and five Brooke rifles.[111]

Dr. James Hazlett, a noted authority on Civil War cannon, has catalogued no less than ninety-two surviving Tredegar field and heavy cannon, some of which are tentative identifications. Doubtless others survive. The Gettysburg National Military Park boasts thirteen Trede-

RESTORATION OF TREDEGAR

Tredegar was in operation until modern times but has been essentially deserted since 1957. The Ethyl Corporation of Richmond, present owners of the site, are now in the process of restoring a portion of the Tredegar plant. The reconstructed chimney attached to the gun foundry was completed in 1975.

(The Ethyl Corporation)

gar cannon—two bronze 12-pounder howitzers, three bronze 12-pounder Napoleons, one 3-inch rifle, one 10-pounder Parrott, and six 20-pounder Parrotts. The Shiloh park has several excellent examples of bronze 6-pounders and 12-pounder howitzers. An 1863 10-pounder Parrott is also at the park. Manassas National Battlefield Park has one 3-inch rifle and three (two tentatively identified) iron 6-pounders. The only surviving bronze Tredegar 12-pounder mountain howitzer and one of only two surviving iron 12-pounder Napoleons is in the possession of one of the authors. The Petersburg National Battlefield Park has the other iron Napoleon as well as several iron 12-pounder howitzers. The Washington Navy Yard has an excellent collection of Tredegar Brookes.

As a rule Tredegar field pieces are stamped on the right trunnion face with the firm initials: "J.R.A. & CO. T.F." (Joseph R. Anderson & Company Tredegar Foundry). The year of casting is on the left trunnion face and the foundry number is on the muzzleface. There are some variants to this, however. Foundry numbers are occasionally not inscribed on some guns. A bronze 6-pounder at the Kennesaw Mountain National Battlefield Park has a unique "TREDEGAR" and "RICHMOND, VIRGINIA" marking. The weapon was cast prior to the war. Inspector's initials are occasionally found on some guns.

NOTES

[1] Dew, *Ironmaker to the Confederacy,* 2-4, 13-16, 19-20.

[2] *Ibid.,* 11-12.

[3] *Ibid.,* 69-79; Account Book, 1861-63, of Edward R. Archer, Virginia Historical Society. Hereafter cited as Archer Account Book, VHS.

[4] Dew, *Ironmaker to the Confederacy,* 83-85.

[5] *Ibid.,* 110, 114; Josiah Gorgas, "Notes on the Ordnance Department of the Confederate Government," *Southern Historical Society Papers,* XII (1884), 71-72; Richmond *Dispatch,* October 3, 1861; Richmond *Enquirer,* October 1, 1861.

[6] E. P. Alexander, *Military Memoirs of a Confederate* (Bloomington: Indiana University Press, 1962), 24.

[7] Jennings C. Wise, *The Long Arm of Lee* (New York: Oxford University Press, 1959), 65.

[8] Susan P. Lee, ed., *The Memoirs of William Nelson Pendleton* (Philadelphia: J. B. Lipponcott Company, 1893).

[9] Gorgas, "Notes on the Ordnance Department," 75.

[10] J. R. Anderson & Company, Papers Relating to Citizens or Business Firms, National Archives Record Group 109 (War Department's Collection of Confederate Records). Hereafter cited as "Citizens File."

[11] Dew, *Ironmaker to the Confederacy,* 108.

[12] *Ibid.,* 112.

[13] Richmond *Enquirer,* September 28, 1861.

[14] Dew, *Ironmaker to the Confederacy,* 112.

[15] *Ibid.,* 109.

[16] *Ibid.,* 118-119; Ripley, *Artillery and Ammunition of the Civil War,* 371.

[17] Tredegar Gun Foundry Book 1861-65, Virginia State Library. Hereafter cited as Tredegar Gun Foundry Book, VSL.

[18] Richmond *Enquirer,* September 28, 1861; Dew, *Ironmaker to the Confederacy,* 100-101, 135-136; Kathleen Bruce, *Virginia Iron Manufacture in the Slave Era* (N.Y.: The Century Company, 1930), 192. A Tredegar gun cast of Graham iron also burst at Columbus, Kentucky.

[19] *Official Records of the Union and Confederate Navies in the War of the Rebellion* (Washington: Government Printing Office, 1894-1927), Series II, I, 718-719. Hereafter cited as *O.N.R.*

[20] *O.R.,* XI, pt. 3, 461.

[21] Dew, *Ironmaker to the Confederacy,* 180.

[22] J. R. Anderson & Company, "Citizens File."

[23] Archer Account Book, VHS; Tredegar Gun Foundry Book, VSL; Dew, *Ironmaker to the Confederacy,* 97.

[24] Ripley, *Artillery and Ammunition of the Civil War,* 22-24, 46.

[25] Richmond *Dispatch,* July 8, 1900; "Historical Sketch of the Rockbridge Artillery, C.S. Army," *Southern Historical Society Papers,* XXIII (1895), 129, 130, 135.

[26] Dew, *Ironmaker to the Confederacy,* 181.

[27] Tredegar Gun Foundry Book, VSL.

[28] J. R. Anderson & Company, "Citizens File."

[29] Dew, *Ironmaker to the Confederacy,* 123.

[30] Tredegar Gun Foundry Book, VSL.

[31] Philadelphia *Inquirer,* June 7, 1862.

[32] Memphis *Appeal,* April 12, 1862.

[33] David G. McIntosh, "The Confederate Artillery—Its Organization and Development," *The Photographic History of the Civil War* (New York and London: Thomas Yoseloff, 1957), V, 56.

[34] Dew, *Ironmaker to the Confederacy,* 123; Wise, *The Long Arm of Lee,* 32-33. Two six gun batteries of Williams breech-loaders were made at Lynchburg and one at Mobile. The latter was used by Buckner. Samson & Pae of Richmond also produced four of these weapons.

[35] Columbus *Enquirer,* May 13, 1862.

[36] *O.R.,* XI, pt. 3, 583.

[37] Wise, *The Long Arm of Lee,* 65.

[38] Edwin C. Bearss, "Artillery Study Shiloh National Military Park," Project No. 17.

[39] L. Van Loan Naisawald, *Grape and Canister* (New York: Oxford University Press, 1960), 35-37.

[40] James C. Hazlett, "The Federal Napoleon," *The Military Collector and Historian,* XV (Winter 1963), 104; U.S. House of Representatives Document No. 99, 2nd Session, 40th Congress, 1867-68 (Washington: Government Printing Office, 1868), 864-881.

[41] James C. Hazlett, "The Napoleon Gun, Its Origin and Introduction into American Service," *The Military Collector and Historian,* XV (Spring 1963), 1-2.

[42] Gorgas, "Notes on the Ordnance Department," 93.

[43] Ripley, *Artillery and Ammunition of the Civil War,* 109-110, 370.

[44] James C. Hazlett, "The Parrott Rifles," *Civil War Times Illustrated,* V (November 1966), 29-30; Wise, *The Long Arm of Lee,* 63-64.

[45] Tredegar Gun Foundry Book, VSL.

[46] E. P. Alexander, "Confederate Artillery Service," *Southern Historical Society Papers,* XI (1883), 104-107.

[47] Fairfax Downey, *The Guns at Gettysburg* (New York: David McKay Company, Inc., 1958), 97.

[48] James C. Hazlett, "The Confederate Napoleon Gun," *The Military Collector and Historian,* XVI (Winter 1964), 108-110.

[49] Columbus *Enquirer,* July 7, 1863.

[50] Downey, *The Guns at Gettysburg,* 97.

[51] Wise, *The Long Arm of Lee,* 370.

[52] Tredegar Foundry Sales Book, 1861-67, Virginia State Library. Hereafter cited as Tredegar Foundry Sales Book, VSL.

[53] Wise, *The Long Arm of Lee,* 72.

[54] *O.R.,* XI, pt. 2, 510-513, 551, 726.

[55] *Ibid.,* XII, pt. 2, 558.

[56] Wise, *The Long Arm of Lee,* 284-286.

[57] *Ibid.,* 298-299.

[58] Ralph W. Donelly, "Confederate Copper," *Civil War History,* I (Dec. 1955), 359.

[59] Dew, *Ironmaker to the Confederacy,* 192-193.

[60] *O.R.,* XXI, 1046-1047.

[61] *Ibid.*

[62] Tredegar Foundry Sales Book, VSL.

[63] Dew, *Ironmaker to the Confederacy,* 187.

[64] Bell I. Wiley, ed., *Recollections of a Confederate Staff Officer by G. Moxley Sorrel* (Jackson, Tenn.: McCowat-Mercer Press, Inc., 1958), 134.

[65] Douglas S. Freeman, ed., *A Calendar of Confederate Papers* (Richmond: The Confederate Museum, 1908), 323.

[66] Dew, *Ironmaker to the Confederacy,* 188-189.

[67] J. R. Anderson & Company, "Citizens File."

[68] *O.R.,* XXV, pt. 2, 617-618.

[69] *Ibid.,* 635-638.

[70] *Ibid.,* 749.

[71] Dew, *Ironmaker to the Confederacy,* 190.

[72] Tredegar Gun Foundry Book, VSL.

[73] Tredegar Foundry Sales Book, VSL; Wise, *The Long Arm of Lee,* 563, 570-571; *O.R.,* XXV, pt. 2, 838-839.

[74] Dew, *Ironmaker to the Confederacy,* 190-191.

[75] *Ibid.,* 193; *O.R.,* XXV, pt. 2, 749; Tredegar Gun Foundry Book, VSL.

[76] Richmond *Examiner,* May 16, 1863; Dew, *Ironmaker to the Confederacy,* 193–194.

[77] *Ibid.,* 195.

[78] Wise, *The Long Arm of Lee,* 666–668; Downey, *The Guns at Gettysburg,* 133. Dew theorizes that the lack of Tredegar Parrotts significantly affected the battle of Gettysburg. See Dew, 195, 288.

[79] *O.R.,* LIII, 287.

[80] S. H. Lockett, "The Defense of Vicksburg," *Battles and Leaders of the Civil War,* III (New York and London: Thomas Yoseloff, 1956), 483; Edwin C. Bearss and Warren E. Grabau, "How Porter's Flotilla Ran the Gauntlet Past Vicksburg," *Civil War Times Illustrated,* I (Dec. 1962), 41–42, 48.

[81] *O.R.,* XXIV, pt. 2, 337, 339, 341.

[82] Dew, *Ironmaker to the Confederacy,* 277; *O.N.R.,* Series 2, II,, 548, 641, 757.

[83] *Ibid.,* 550.

[84] Tredegar Gun Foundry Book, VSL.

[85] *O.R.,* XIV, 865.

[86] Tredegar Foundry Sales Book, VSL.

[87] *O.N.R.,* Series 2, II, 418.

[88] *O.R.,* XVIII, pt. 2, 234.

[89] *Ibid.,* 220.

[90] *Ibid.,* 346.

[91] *Ibid.,* 364 and XXV, pt. 1, 514.

[92] Dew, *Ironmaker to the Confederacy,* 276–277.

[93] U.S. House of Representatives Document No. 99, 2nd Session, 40th Congress, 1867–1868 (Washington: Government Printing Office, 1868), 711–714, 805–812.

[94] Tredegar Gun Foundry Book, VSL.

[95] Listing of Cannon in Walker's Division, Army of Mississippi, 1863 (month unknown), Mississippi State Archives.

[96] *O.R.,* XXX, pt. 2, 41–42.

[97] Record of Receipts and Issues at the Atlanta Armory, Dec. 1863–June 1864, Vols. 17–18, National Archives Record Group 109 (War Department's Collection of Confederate Records). Information courtesy of Edwin Olmstead.

[98] Ripley, *Artillery and Ammunition of the Civil War,* 28.

[99] Gorgas, "Notes on the Ordnance Department," 93.

[100] Tredegar Gun Foundry Book, VSL.

[101] Dew, *Ironmaker to the Confederacy,* 279.

[102] Tredegar Gun Foundry Book, VSL.

[103] As quoted in Bruce, *Virginia Iron Manufacture in the Slave Era,* 337.

[104] Ripley, *Artillery and Ammunition of the Civil War,* 56–60; Maury Klein, *Edward Porter Alexander* (Athens, Ga.: University of Georgia Press, 1971), 121–122; Dew, *Ironmaker to the Confederacy,* 279.

[105] *O.R.,* XLIII, pt. 1, 58.

[106] *Ibid.,* XLVI, pt. 1, 1073–1074.

[107] Dew, *Ironmaker to the Confederacy,* 238, 247, 288; William LeRoy Broun, "The Red Artillery," *Southern Historical Society Papers,* XXVI (1898), 374; *O.R.,* Series 4, III, 734.

[108] James H. Brewer, *The Confederate Negro: Virginia Craftsmen and Military Laborers, 1861–1865* (Durham, N.C.: Duke University Press, 1969), 60.

[109] Dew, *Ironmaker to the Confederacy,* 287.

[110] *Ibid.,* 285.

[111] *O.R.,* XLVI, pt. 3, 1009–1010.

Present day ruins of the Bellona Foundry. *(Edwin Olmstead)*

BELLONA FOUNDRY

The Bellona Foundry, established in 1816 by Major John Clark, was located on the James River in Chesterfield County, Virginia, thirteen miles upriver from Richmond. In the early 1840's Clark sold the establishment to Dr. Junius L. Archer. Dr. J. L. Archer is sometimes confused with Dr. Robert Archer, father-in-law of Joseph R. Anderson, partner in the Tredegar Iron Works, and probable inventor of the Archer projectile, used by the Confederates early in the war. Under both John Clark and Junius Archer, Bellona became an important supplier of cannon to the Federal Government. Close to the foundry Federal authorities set up an arsenal also known as Bellona, which repaired small arms, fabricated munitions, and stored cannon cast at the foundry.[1]

Archer, like Anderson at Tredegar, was against the Rodman method of casting when introduced in the 1840's. In 1859 the Ordnance Department refused to accept any further cannon not cast by this method, thus temporarily ending Bellona's ordnance trade with the Federal Government. A reprieve came in February, 1860, however, when Secretary of War John B. Floyd permitted Bellona to cast by the old method forty-six cannon remaining from an 1857 order.[2]

Bellona was thus completing the order when war clouds began to loom over the country. By April, 1861, about twenty-three guns of various calibers had been completed and ordered shipped to Ft. Monroe by the Ordnance Department. The Virginia State Assembly took immediate action to insure that the guns would not be removed. The weapons were confiscated "for safe keeping" and the Federal Government reimbursed.[3] Thus, the State of Virginia came into possession of a number of Bellona cannon on the eve of the war.

Joseph Anderson rented Bellona in early 1861 but

Bellona gun flask and unfinished cannon thrown into James River by Federal raiders and now surviving on the Bellona grounds. *(Edwin Olmstead)*

soon abandoned it in favor of expanding his own works.[4] By August the Confederate Government had contracted with the firm for 120 field pieces, consisting of howitzers and 3-inch rifles.[5] No Confederate Bellona field guns are known to survive. The firm soon became heavily engaged in the casting of heavy guns and the probabilities are that the order for field guns went unfilled.

There were some 3-inch rifles in use which the Confederates termed "Burton and Archers." In November, 1861, the 2nd Company of the Louisiana Washington Artillery was reported to possess four Burton and Archer bronze 3-inch rifles.[6] On May 20, 1862, the Rowan (N.C.) Artillery was issued two Burton and Archer iron 3-inch rifles. One was swapped for a Union 3-inch rifle at Gaines Mill and the other burst at Gettysburg.[7] In February, 1863, the 2nd Corps artillery of Lee's Army listed several Burton and Archers in use.[8] In all liklihood these guns were not cast by Archer at Bellona. The term "Burton and Archer" probably referred to the fact that the guns were rifled to fire Burton and Archer projectiles.

Although overshadowed by her sister foundry downriver, Bellona none-the-less proved to be of primary importance to the Confederacy. During 1862 Bellona delivered about thirty-seven heavy cannon—fifteen 8 and 10-inch columbiads, three 24-pounders, two 9-inch guns, five 4.2-inch siege guns, and two 7-inch guns. Tredegar did finishing work on some of the Bellona castings, including rifling, sighting, and sometimes reboring.[9]

After casting, the guns were placed on barges and sent downriver to Richmond where they were unloaded and test fired. The weapons were then reloaded on railroad cars for such destinations as Charleston, Savannah, and Mobile. All of this work was done by Tredegar laborers. Loading these huge weapons often proved a difficult task. On November 12, 1863, for example, a 10-inch Bellona smoothbore was placed on a flatcar at the Richmond Depot, and the sheer weight of the gun broke the car down.[10]

In April, 1863, a fire seriously damaged a large portion of the foundry. The Government turned to the Tredegar works to re-equip Bellona but Anderson vigorously advised against it. He argued that there were insufficient raw materials to keep Tredegar operating at full capacity.[11] Anderson was overruled, however, and reconstruction was begun. In June the Government impressed a large lathe which had been recently purchased by Tredegar at Raleigh, North Carolina, and, much to the distaste of Anderson, diverted the machine to Bellona.[12]

Production was apparently not halted for long. In June a Richmond correspondent reported that "the 10-inch columbiads now being manufactured at the Bellona Arsenal, near this city, are said to be equal to any manufactured in this country. These works have been turning out guns from the commencement of the war to the pres-

Top: An 8-inch Bellona columbiad at the Vicksburg National Military Park.
Middle: 9-inch Dahlgren cast by Bellona in 1862 and surviving at Vicksburg.
Below: Right trunnion face marking on a Bellona gun. The inscription stands for "Bellona Foundry Junius L. Archer".

(Tarry Beasley)

An 1864 Bellona 8-inch columbiad in the West Point Collection. *(West Point Museum)*

ent and not the first one has proved defective in any respect."[13]

Despite public reassurance, the April fire, coupled with iron shortages, inevitably took a heavy toll on production. From an 1862 delivery of thirty-seven guns, the 1863 delivery dropped to only sixteen guns. These included two 7-inch guns, two 9-inch guns, and twelve 10-inch columbiads.[14]

How many guns were cast for the Confederate Government is purely conjectural, but an estimation can be made based on the foundry numbers of surviving weapons. Eleven Confederate Bellona guns are known to survive, one of which is a tentative identification. The types and numbers of surviving weapons are as follows:[15]

TYPE	YEAR	PRESENT LOCATION	NUMBER
8-in.	1861	St. Augustine*	27
8-in.	1861	St. Augustine*	29
9-in.	1862	Vicksburg	**
10-in.	1862	Vicksburg	**
10-in.	1862	Charleston	**
8-in.	1862	Fort Darling, Va.	66
10-in.	1863	Vicksburg	13
10-in.	1863	Charleston	20
10-in.	1863	Vicksburg***	**
10-in.	1864	Vicksburg	46
10-in.	1864	West Point	67

*Both 1861 guns have a "C.S." inscription.

**Indicates information either not marked or unreadable.

***Tentative identification.

For some unknown reason a new series of foundry numbers seems to have been started in late 1862. This can be substantiated not only from the foundry numbers of surviving weapons but also in the Tredegar Foundry records. In December, 1862, Tredegar laborers hauled a 9-inch Bellona gun which was entered as being "NO. 6." The only other instance in which a foundry number was recorded was on April 11, 1863, when two 10-inch Bellona smoothbores, numbers 11 and 12, were shipped to Jackson, Mississippi.[16] From all available sources, then, it can be determined that Bellona cast in the neighborhood of 135 heavy guns for the Confederate Government, second only to the Tredegar works.

Two of the last few guns cast for the United States survive and probably both were in the lot confiscated by the State of Virginia. One, an 1860 42-pounder, is at the Washington Navy Yard. The other, an 1861 24-pounder howitzer, is in private hands. Both are stamped "U.S."

Early Bellona guns were stamped either "I.C. B.F" or "J.C. B.F." (John Clark Bellona Foundry). A March, 1863, list of the guns at Port Hudson, for example, records two guns, 42-pounders, inscribed in such a manner. Although the dates were not recorded, the foundry numbers were 163 and 183.[17] Surviving Confederate guns are quite often marked with a "C.S." or an "R.M.C." (Richard M. Cuyler). Cuyler, commanding the Macon Arsenal, inspected many of the guns. The guns are usually inscribed "B.F. J.L.A." (Bellona Foundry Junius L. Archer). A surviving 1864 10-inch columbiad, number 67, is inscribed "B.A." (Bellona Arsenal) and "R.M.C."

NOTES

[1] Bruce, *Virginia Iron Manufacture in the Slave Era,* 116, 192; Ripley, *Artillery and Ammunition of the Civil War,* 361; Richmond *Enquirer,* September 18, 1816. Between 1826 and 1838 Bellona furnished 204 24 pounder seige guns, 446 32 pounders and seventeen 42 pounders. ((Tabular reports of the inspection of cannon made in various foundries 1826–1838, National Archives Record Group 156, entry 210. Information courtesy of Edwin Olmsted.)

[2] Dew, *Ironmaker to the Confederacy,* 47.

[3] *O.R.,* Series 4, I, 203–204; Ripley, *Artillery and Ammunition of the Civil War,* 40–41.

[4] Dew, *Ironmaker to the Confederacy,* 108.

[5] Lee, ed., *The Memoirs of William Nelson Pendleton,* 167–168.

[6] New Orleans *Picayune,* November 10, 1861.

[7] Walter Clark, *Histories of the Several Regiments and Batteires From North Carolina in the Great War 1861–1865* (Raleigh: State of North Carolina, 1901), Vol. I, 566, 568. The authors' thanks to Edwin Olmstead for calling our attention to this reference.

[8] *O.R.,* XXV, pt. 2, 635–638.

[9] Tredegar Foundry Sales Book, VSL.

[10] *Ibid.*

[11] Bruce, *Virginia Iron Manufacture in the Slave Era,* 414–415.

[12] Dew, *Ironmaker to the Confederacy,* 199.

[13] Richmond *Dispatch,* June 14, 1863.

[14] Tredegar Foundry Sales Book, VSL.

[15] Based on "Bellona Digest and Index," an unpublished paper by Edwin Olmstead.

[16] Tredegar Foundry Sales Book, VSL.

[17] *O.R.,* XV, 1027.

THE IMPROVISED FOUNDRIES

BENNETT & LURGES

One of several New Orleans foundries which expressed an early desire to cast heavy seacoast cannon was the firm of Bennett & Lurges, also known as the New Orleans Foundry & Ornamental Works. The facility, under the proprietorship of B. K. T. Bennett and Francis Lurges, was located at the corner of Magnolia and Erato Streets. Pre-war products included railings, ornamental verandahs, and iron work of every description.[1]

The Bennett & Lurges firm was most anxious to engage in the casting of heavy guns under a Confederate contract. They had earlier completed a heavy gun which was mounted at Columbus, Kentucky, and Island Number 10. The owners later complained that they had been willing to turn out guns from the beginning, "but had received no encouragement from the government." [2]

The Navy Department eventually contracted with the firm for all the 8-inch columbiads they could make, without stipulation as to price, until they were ready to make larger cannon. The models for 9-inch guns were in a forward state when the city fell. Since the foundry had only one lathe for the turning of heavy guns, a lathe at the machine shop of the Jackson & Great Northern Railroad and another at the Shakespeare Foundry were placed at their disposal.[3]

By September, 1861, work had been completed on the first Navy columbiad. "An 8-inch columbiad has been manufactured at the foundry of Messrs. Bennett & Lurges of New Orleans under the superintendence of Mr. Daniel Brasill, according to the most approved pattern," reported a Natchez newspaper. "It was cast solid and bored. The time occupied in boring was two weeks, and when it was tested, according to the general rules for that purpose, it was found to stand 'all that could be put upon it.' Shell were thrown from it without straining, a distance of full two miles." [4]

At the time the city fell the firm had cast five 8-inch columbiads weighing 11,000 pounds each. Three of the guns were completed, two of which were mounted on the floating battery below the city and one at Fort St. Phillip. One officer later reported that Bennett & Lurges "made the most reliable guns of any foundry in New Orleans." [5]

Some work was also done on field guns. On February 1, 1862, a 6-pounder bronze smoothbore was delivered. On March 12th six iron 6-pounders, costing $300 apiece, were shipped.[6] A Navy Department contract for twelve bronze 6-pounder rifles apparently failed to materialize.[7] Several field pieces and two unfinished columbiads were left at the foundry upon the evacuation of the city.[8] Besides cannon the establishment also delivered over 4000 rounds of shot and shell.[9]

There are no known surviving weapons.

NOTES

[1] New Orleans *Picayune,* September 10, 1861; William A. Albaugh and Edward N. Simmons, *Confederate Arms* (Harrisburg, Pa.: The Stackpole Company, 1957), 202, 205.

[2] *O.R.,* VI, 576.

[3] *Ibid.,* 577, 599.

[4] Natchez *Courier,* September 27, 1861. See also Memphis *Appeal,* October 2, 1861.

[5] *O.N.R.,* Series 2, I, 561.

[6] Bennett & Lurges, "Citizens File."

[7] *O.N.R.,* Series 1, XVII, 163.

[8] *O.R.,* VI, 576, 584.

[9] Bennett & Lurges, "Citizens File."

T. M. BRENNAN & COMPANY

By the fall of 1861, the city of Nashville, Tennessee, had become the most important logistics center in the Confederacy west of the Alleghanies. Reports soon filtered to the North concerning the war industrial activity developing in that city. One such report mentioned a cannon foundry where 100 men were employed day and night in the casting of cannon, shot and shell.[1] A later, similar report also mentioned a large foundry which "turned out a large amount of iron light artillery of every description." [2] These accounts were references to the foundry of T. M. Brennan & Company.

The Brennan foundry, also known as the Claiborne Machine Works, was located on Front Street near Broad. The plant consisted of an extensive foundry and machine shop and prior to the war engaged in the manufacture of machinery, steam engines, boilers, saw mills, mill gearing, and ornamental castings. Some 100 hands were employed. Thomas Brennan was a former mechanic who had been engaged in the foundry business in Nashville for a number of years.[3]

On May 16, 1861, the State of Tennessee contracted with the firm for two batteries of iron cannon, each consisting of four 6-pounders and two 12-pounder howitzers. The contract price was $.15 per pound. A contract for an additional battery of guns, four 6-pounders and two 12-pounder howitzers, was granted on May 21st. Shot and shell were also to be furnished at the same contract price as cannon. The contracts were signed on behalf of the firm by M. G. L. Claiborne. On May 25th another State contract was made for battery equipment (carriages, caissons, limbers) for six guns. By August 31st the State had paid the firm $17,639.99. Indications are that all guns and equipment were delivered.[4]

In July, 1861, a correspondent of the Nashville *Union and American* visited the plant and found preparations well underway. He reported:

> NASHVILLE MADE CANNON.—We saw yesterday at the Foundry of Mr. T. M. Brennan, in this city, a battery of cannon at his establishment. The cannon were manufactured of iron from the works of Messrs. Woods, Yeatman & Co. and Hillman Brothers, and from the severe tests to which they were subjected, the iron proves to be admirably adopted to the manufacture of heavy guns. These cannon weigh only about ten pounds more than brass guns of the same calibre. Mr. Brennan is making arrangements to turn out a complete battery each week, the casting, woodwork, and everything pertaining being done at his establishment. Mr. Brennan is also making quantities of shot and shell.[5]

One of the first units to receive Brennan cannon was Rutledge's (Tennessee) Battery. The armament consisted of four 6-pounders and two 12-pounder howitzers. A picture in Miller's *Photographic History of the Civil War* (Vol. V, 65) depicts a group of men from this Battery gathered around what could well be an early Brennan fieldpiece. Unfortunately, only the muzzle is visible.

Brennan specialized in iron ordnance. ". . . I am now manufacturing a very superior cast iron gun from the best Tennessee cold blast iron thoroughly mixed and manipulated in an air furnace put up expressly for the manufacture of guns and combining all the recent improvements," wrote Brennan to General Leonidas Polk in September. "This gun has been successfully tested with a quadruple service charge. It is a new and highly approved pattern, accurately bored out, and finished on the exterior, and does not exceed in weight the average of bronze field guns." [6]

By November Brennan had begun monthly deliveries of cannon to the C. S. Government. At that time his factory had the capacity to turn out twelve field pieces, 6-pounders and 12-pounder howitzers, and three siege guns, up to 32-pounder, per week. Ten tons of shot and shell could be cast daily.[7]

Despite increased production, heavy requisitions for cannon continued to strain the Ordnance Department's capability to supply. On October 6th General Simon Buckner at Bowling Green requested six 12-pounder siege guns, six 12-pounder howitzers, and a few rifled guns to complete the defenses at that place. The Nashville Arsenal was unable to fill the order, however, and the General was forced to go to Memphis for the weapons.[8] A few days later Samuel Tate, aide to General Albert Sidney Johnston, reported that "the foundries of Nashville are prepared to work on a large scale, but have engagements for some time ahead; but in case of emergancy I have no doubt you can get the guns they are making for the Government." [9]

Chief Engineer J. F. Gilmer estimated that twenty to twenty-five 12-pounder siege guns were needed to protect the land defenses around Nashville.[10] In early January, 1862, Lieutenant M. H. Wright, commanded the Nashville Arsenal, contracted with Brennan for twelve batteries of 12-pounder siege guns, complete with carriages, caissons, and 5000 rounds of shot and shell. Only a few 12-pounders were delivered before the fall of the city in February, 1862.[11]

Between November, 1861, and February, 1862, Brennan made contract deliveries to the Confederate Government as follows:[12]

DATE	GUNS DELIVERED	ARSENAL
Nov. 9, 1861	2 6-pounders 4 6-pounders (unfinished) 4 12-pdr. howitzers	Nashville

The T. M. Brennan Cannon Foundry, Nashville, Tennessee, from a photograph taken by Federal soldiers in 1864. *(National Archives)*

Markings on this 1861 iron 6-pounder field gun indicate that it was cast by the Brennan foundry for the State of Tennessee. It survives at the Vicksburg National Military Park. *(Tarry Beasley)*

	2 12-pdr. howitzers (unfinished)	
	1 12-pounder	
Nov. 30	6 6-pounders	"
	2 6-pounders (unfinished)	
	4 12-pdr. howitzers	
	6 Guns (caliber unspecified)	
Dec. 31	4 6-pounders	"
	2 12-pdr. howitzers	
	8 Guns (caliber unspecified)	
Jan. 31,1862	8 6-pounders	"
	4 12-pdr. howitzers	
	4 6-pounders	"
Feb.	4 12-pdr. howitzers	
	3 12-pounders	
	3 12-pounders (unfinished)	
	TOTAL DELIVERED 71	

Some of the unfinished castings were sent to a Clarksville, Tennessee, foundry for completion. Brennan was apparently unable to finish the guns fast enough to keep up with the casting of the tubes. Although not mentioned by name, this undoubtedly is a reference to Whitfield, Bradley & Company of Clarksville. The records of that company are sketchy, but a voucher does reveal that the firm was paid money on January 2, 1862, for "boring out, turning up, dressing & finishing" four 12-pounder howitzers and two 6-pounders. Also, for February 12 for "turning and fitting" four 12-pounder howitzers and two 6-pounders the firm was paid $600.00.[13]

Since all the cannon supplied by Brennan were of iron, the cost was only $250 apiece for the howitzers and 6-pounders, $400 apiece for the 12-pounders, and a correspondingly reduced cost for the unfinished castings. Brennan also supplied thousands of rounds of shot and shell, seventeen gun carriages, twelve caissons, two forges, two battery wagons, harness, and a variety of other items.[14]

In his September communication to Polk, Brennan mentioned that he was "getting up patterns for a very complete rifling machine which will rifle guns of any caliber." Just before the evacuation of Nashville, Brennan did complete a very expensive and valuable rifling machine. A spy was sent north to the Fort Pitt Foundry where he obtained the proper drawings and specifications which enabled Brennan's mechanics to put up the machine. It was later removed to Atlanta.[15]

Nashville fell to the advancing Union forces in February, 1862. About fourteen Brennan cannon, some of them unfinished, were hurriedly evacuated to Atlanta. Apparently not all guns were salvaged. Several weeks after the capture of the city, an eyewitness reported seeing "about a dozen large cannon still lying at this foundry, which the sudden flight of the rebels prevented them from rifling and carrying away." [16] Brennan was arrested by Federal authorities and tried for treason for his role in Confederate arms production.[17]

A few specific Brennan weapons traceable through records and surviving cannon are as follows:[18]

WEAPON	NUMBER	REMARKS
6-pounder	5	Shipped from Atlanta Arsenal on Aug. 23, 1863, to Etowah, Ga.
6-pounder	11	Survives at Chalmette National Cemetery in New Orleans
6-pounder	28	Shipped through Atlanta Arsenal to Augusta Arsenal on June 29, 1864
6-pounder	46	Survives at Fort Donelson
6-pounder	59	Shipped through Atlanta Arsenal to Augusta Arsenal on June 29, 1864
6-pounder	—	Survives at Vicksburg. Marked "1861" and "TENN."
6-pounder	—	Turned into Atlanta Arsenal on June 6, 1864
12-pdr. Howitzer	—	Survives in private hands. Marked "1861"

Surviving weapons are marked on the right trunnion face: T. M. BRENNAN MAKER NASHVILLE, TENN." The surviving gun at Vicksburg is also marked "TENN.", indicating that the gun was made under State contract. An "M.H.W." is occasionally stamped, which indicates that the weapon was inspected by "Moses Hannibal Wright" of the Nashville Arsenal. Brennan 6-pounders and 12-pounder howitzers are very plain and bear no resemblance to the standard U.S. model 1841 pieces of the same caliber.

NOTES

[1] New York *Times,* October 18, 1861.

[2] Walter G. Stevenson, *Thirteen Months in the Rebel Army* (New York: Barnes, 1959), 61.

[3] *Nashville City and Business Directory for 1860-61,* 67-278.

[4] Military Board Record Book, Vol. 63, pp. 17, 21, 24, 75, Series No. 6, Record Group 21, Tennessee State Archives.

[5] Nashville *Union and American,* July 19, 1861.

[6] T. M. Brennan, "Citizens File."

[7] *O.R.,* IV, 481.

[8] *Ibid.,* 438.

[9] *Ibid.,* 512.

[10] *Ibid.,* 514.

[11] T. M. Brennan, "Citizens File."

[12] *Ibid.*

[13] Whitfield, Bradley & Company, Ibid.

[14] T. M. Brennan, Ibid.

[15] Stevenson, *Thirteen Months in the Rebel Army,* 61.

[16] *Ibid.*

[17] Columbus *Enquirer,* April 22, 1862.

[18] Record of Receipts and Issues at the Atlanta Armory, December, 1863-June, 1864, chap. IV, Vol. 18, Pps 82, 164, National Archives Record Group 109, War Department's Collection of Confederate Records. (Information courtesy of Sydney Kerksis and Edwin Olmstead.)

BUJAC & BENNETT

About November, 1861, B. K. Bennett, one of the principals in the Bennett & Lurges firm of New Orleans, formed a partnership with M. J. Bujac, part owner of the rifle and sporting apparatus firm of Bujac & Hensler.[1] The original purpose of the partnership was to engage in the manufacture of small arms in New Orleans, but the firm later geared to cannon production. By their individual efforts air furnaces were built and six lathes were obtained for the production of heavy seacoast guns.[2]

Bujac & Bennett eventually cast six 32-pounder seacoast guns for the Navy Department, but casting methods were so crude that all failed to pass inspection. The exteriors and the bores of the guns were filled with cavities. "The cavities on the exterior should not exceed one-tenth of an inch, but the holes on the inside were much deeper than that," described the inspector. Due to the shortage of heavy guns some of the weapons were later accepted, but, warned the inspector, "there was no telling when they might burst." [3]

The difficulty caused over the condemned guns resulted in a rift between the owners and the Navy Department. The owners, thereafter, positively refused to work for the Navy, although they had one of the most extensive establishments in the city and a large quantity of the right kind of iron. They complained that the Navy was too particular in their inspections.[4]

The firm also experienced difficulties in the casting of field guns. In December, 1861, a dozen "Parrott rifles" were delivered, three of which later burst. On March 8, 1862, eight more Parrotts were delivered at $235 apiece.[5]

On the evacuation of New Orleans the firm removed the rifle machinery for small arms with a plan to reopen operations in Montgomery, Alabama. The machinery was later sold to the Alabama Arms Company of that city.[6]

There are no known surviving cannon.

NOTES

[1] Albaugh and Simmons, *Confederate Arms,* 205.
[2] *O.R.,* VI, 576, 584.
[3] *O.N.R.,* Series 2, I, 526, 561.
[4] *Ibid.,* 468.
[5] M. J. Bujac, "Citizens File."
[6] Albaugh and Simmons, *Confederate Arms,* 205.

JOHN CLARK & COMPANY

The New Orleans foundry of John Clark & Company, located at the corner of Race and Tchoupitoulas Streets, had been in operation only a few years when the country became engulfed in Civil War. Such peacetime products as furnaces, grate bars, car axles, and cotton presses gave way to the manufacture of cannon and ship machinery.[1]

By the summer of 1861 the firm was engaged in the casting of field guns, 6-pounders and 12-pounder howitzers, for private parties. The gun carriages were made at the establishment of Peter Seibert on Liberty Street.[2] In July, 1861, a test was made by a squad of the Washington Artillery on two new 6-pounders cast by the firm. "We understand that the test was perfectly successful," reported the *Picayune,* "the pieces proving to be first rate specimens of ordnance." [3] The weapons cast were very different in outward appearance from the standard Model 1841 pieces of the same caliber, differences probably due to a lack of exact patterns rather than a planned model change.[4]

On March 11, 1862, Clark addressed General P. G. T. Beauregard at his Jackson, Tennessee, headquarters regarding the activities at the plant. "We beg leave to advise you that we are prepared to turn out six guns per week complete, provided we have the material," he stated. "We want copper, block tin or bells." At that time the firm had already furnished over 100 cannon to various units, including the Washington Artillery, Watson's (Louisiana) Battery, and the Orleans Guard Artillery.[5] Clark was paid a large advance by the Confederate Government to produce guns on the English Armstrong pattern, but the city was captured before any casting was commenced.[6]

Considering the large number of guns cast, surviving records on the Clark foundry prove most disappointing. Since all weapons were apparently cast under private contract, there is no specific breakdown available as to

A John Clark & Company 6-pounder field gun in the West Point Collection. *(West Point Museum)*

Top: Right trunnion face inscription on a Clark 12-pounder howtizer. *(Tarry Beasley)*
Below: A Clark 12-pounder howitzer at Shiloh.

types cast and dates of delivery. Indications are that most were 6-pounder guns and 12-pounder howitzers. An 1863 listing of the guns in Walker's Division, Army of Mississippi, listed one 6-pounder cast by "J. Clark" of New Orleans.[7] At Chickamauga in September, 1863, Confederates recaptured a bronze 12-pounder howitzer made by "J. Clark" of New Orleans.[8]

Surviving weapons include two 6-pounders at West Point. Records indicate that both were captured by Federal troops along the Mississippi River in 1863. Shiloh has a Clark 12-pounder howitzer which is very clearly marked on both trunnion faces: "JOHN CLARK, MAKER, N.O." Apparently not all Clark weapons were stamped with the firm name. Several 12-pounder howitzers are known to survive (one each at Shiloh, Vicksburg, and Petersburg) which are identical to Clark's unusual and distinctive model although they bear no markings. It is assumed that these weapons were also cast in the Clark shops.

NOTES

[1] *Gardner's New Orleans Directory for 1866,* 590; *O.N.R.,* Series 2, I, 503.

[2] New Orleans *Picayune,* June 16, 1861.

[3] *Ibid.,* July 10, 1861.

[4] Ripley, *Artillery and Ammunition of the Civil War,* 22.

[5] John Clark & Company, "Citizens File."

[6] *O.R.,* VI, 627; *O.N.R.,* Series 2, I, 535.

[7] Listing of Cannon in Walker's Division, Army of Mississippi, 1863 (month unknown), Mississippi State Archives.

[8] *O.R.,* XXX, pt. 2, 41.

COLUMBUS IRON WORKS

The Columbus Iron Works of Columbus, Georgia, was founded in 1848 by William Riley Brown and located at the corner of Short and Dillingham Streets. The firm expanded over the years and prior to the Civil War employed over 100 hands. The plant consisted of Brown's Foundry, Golden's Machine Shop, Love's Variety Shop, Churchill's Rolling Mill, Fell's Cannon Foundry, and Stanford's Boiler Manufactory.[1]

Production of cannon began in the summer of 1861. In July two 3-inch rifles were completed for a local battery.[2] In the fall a contract was signed and work begun on a battery of two 12-pounder howitzers, three 6-pounders, and one rifled gun for the Milton (Florida) Artillery.[3] Two bronze 6-pounders, complete with carriages and caissons, were delivered to the Confederate Government in January, 1862.[4]

In February, 1862, a correspondent of the Columbus *Enquirer* visited the plant and proudly reported:

> CANNON MAKING.—We dropped into the extensive establishment of the "Columbus Iron Works," Thursday, to take a look at a number of pieces of ordnance now being finished there. We found brass 12-pounders for throwing round shot, mortars of the same calibre, and wrought-iron rifled cannon. The perfect casting and beautiful finish of these heavy guns must elicit the surprise and admiration of all who see them, and who reflect that Messrs. Stanford, Golden & Co. have but recently turned their attention to this important and difficult enterprise. They have overcome many obstacles, and have made their establishment a valuable and highly useful institution to the Government and an efficient co-worker in the great cause for which the Southern people are struggling.[5]

Shortly thereafter the firm announced that they were prepared to cast six and furnish four cannon per week.[6]

To ease the severe copper shortage local authorities turned to a patriotic populace to supply bells to be melted down for cannon. In July, 1862, it was noted that five large bells were conveyed down Broad Street on a dray destined to be cast into cannon at the Columbus Iron Works.[7] Both of the 6-pounders delivered to the Confederate Government in January, 1862, were made from bell metal.[8] One cannon, a 12-pounder field howitzer, was made from household brasses donated by the ladies of Columbus. The cannon was dubbed "Ladies Defender", which was inscribed near the breech. It survives today at Columbus.

The plant continued production under Army control through June, 1862. "The foundry here is casting bronze guns and is under my control," wrote Captain F. C. Humphreys, commanding the Columbus Arsenal. "I have no howitzers on hand at present, but will have one or two in a day or more." [9] In June the facility was leased to the Confederate Navy and thereafter used for the manufacture of ship machinery. Under naval control the plant expanded, additional land was leased, and the work force increased to 300 hands.[10] Army ordnance production after that date was shifted to the Columbus Arsenal.

At least one cannon was turned out while the firm was under the Navy Department. The weapon was a 2.75-inch breechloading rifle which survives today at Columbus. It was made in 1863 from the wheelshaft of the sunken river steamer John C. Calhoun. The performance of the weapon was said to have been very poor.[11] In an 1863 listing of the guns in Walker's Division, Army of Mississippi, two 12-pounder howitzers were listed as having been marked "COLUMBUS, GA." [1] It is very possible that these howitzers were cast at the Columbus Iron Works.

NOTES

[1] Diffee W. Standard, *Columbus, Georgia, in the Confederacy* (New York: William-Frederick Press, 1954), 40; "How Our Industry Profits From the War," *Debow's Review* VIII (May-August 1862), 77–78; Albaugh and Simmons, *Confederate Arms,* 211.

[2] Standard, *Columbus, Georgia, in the Confederacy,* 40.

[3] Columbus *Daily Sun,* November 28, 1861.

[4] Columbus Iron Works, "Citizens File."

[5] Columbus *Enquirer,* February 4, 1862.

[6] Memphis *Appeal,* April 12, 1862.

[7] Columbus *Daily Sun,* June 5, 1862.

[8] Columbus Iron Works, "Citizens File."

[9] Humphreys to Cuyler, June 9, 1862, Macon Arsenal Letters Received, Chap. IV, Vol. 4, National Archives Record Group 109 (War Department's Collection of Confederate Records).

[10] William N. Still, "Facilities for the Construction of War Vessels in the Confederacy," *The Journal of Southern History* XXXI (August 1965), 298–299; Memphis *Appeal,* June 12, 1863; "Major James H. Warner," unpublished and unsigned paper in Alabama State Archives; Standard, *Columbus, Georgia, in the Confederacy,* 42–43.

[11] William A. Albaugh, Hugh Benet, and Edward N. Simmons, *Confederate Handguns* (York, Pa.: George Shumway, 1963), 10.

[12] Listing of Cannon in Walker's Division, Army of Mississippi, 1863 (month unknown), Mississippi State Archives.

LEEDS & COMPANY

The largest foundry in New Orleans, Louisiana, was that of Leeds & Company. The establishment was founded in 1823 under the name of C. C. Whitman & Company. In 1825 Mr. Jedediah Leeds acquired one-third partnership and on the following year became the sole owner. Jedediah Leeds died in 1844 and shortly thereafter his two sons, Charles and Thomas, succeeded him. Their cousin, Edward Grinnell, also became a principal in the firm. Thomas Leeds was a draftsman and designer, while Charles took charge of financial matters and Grinnell became superintendent.[2]

Leeds & Company occupied three-fourths of the square of ground bounded by Tchoupitoulas, Delord, Foucher, and St. Joseph Streets. By 1861 the plant had enlarged its work force to 300 hands. The facility produced horizontal and vertical steam engines, boilers, sugar mills, saw mills, gin gearings, "and all machinery required for the South." [2]

Leeds & Company expressed an early desire to cast heavy guns. As early as May, 1861, the firm made inquiries to the Tredegar Iron Works concerning the details of heavy cannon manufacturing. Tredegar authorities replied that they had experienced difficulties primarily in the area of obtaining suitable iron and its proper treatment when obtained.[3] Leeds was soon to share this experience with its Richmond counterpart.

In June, 1861, Leeds produced its first heavy gun, an 8-inch columbiad, for a private party. The weapon received a great deal of attention in New Orleans newspapers and it was anticipated that a large crowd would attend the trial firing.[4] Unfortunately, the weapon failed to pass even the simplest of tests, both poor iron and defective furnaces taking their toll. In tests of this nature a gun standing 800–1000 discharges would be quickly condemned; 1200 discharges would be considered inferior; 1800–2000 would be fair; over 2000 would be excellent. The Leeds columbiad burst on the sixty-third round. The Ordnance Department endorsed the report, despite the owners' complaint that the testing officer had not allowed enough room for recoil, thus causing the bursting.[5]

Despite discouraging initial reports, Leeds proceeded with the casting of heavy guns. The firm eventually produced one heavy gun for the Army and six or eight for the Navy, the first ones being described as "beautiful." Some were placed aboard the naval vessels *Livingston, Ponchatrain,* and *Maurepas* in the river defense fleet, but it was later reported that some of these, like their trial predecessor, also burst.[6]

Casting of field guns got off to a slow start. Rifling machinery was erected at an early date, but relatively few cannon were cast, and those only for private parties.[7] By the end of May, 1861, work had begun on a Saunders Gun, a small caliber brass breechloader with hexagonal bore.[8] In October the first field pieces, four 6-pounders and two 6-pounder rifles, were delivered to the Confederate Government.[9]

In order to expedite the granting of ordnance contracts the Richmond Government often relegated the responsibility to local Department commanders. In the fall of 1861 Major General Braxton Bragg, commanding the Confederate forces between Pensacola and Mobile, began the search for field guns to arm the batteries then forming in his Department. A contract was signed with

LEEDS NAPOLEON

This 1861 Leeds & Company 12-pounder Napoleon was one of six such weapons cast by the firm in 1861. The possibilities are that it served in Robertson's (Florida) Battery during the battle of Shiloh. The gun, at the Petersburg National Battlefield Park, is the earliest known Confederate Napoleon to survive. *(Tarry Beasley)*

Skates & Company of Mobile for four complete field batteries, but the firm later proved unable to deliver.[10] In October Captain H. Oladowski of the Ordnance Department was ordered to furnish two field batteries for the Army at Pensacola. Oladowski ventured far from his area—first to Rome, Georgia, and later to New Orleans. At Rome he found that one battery was already being fitted up for Bragg. He thereupon proceeded to New Orleans and Leeds & Company to arrange a contract. An order was soon placed with the firm for six 12-pounder Napoleons. Contracts were also entered into for three 6-pounders for Major M. L. Smith and five 6-pounder rifles for Captain I. C. Booth.[11]

Work progressed rapidly on the order. By December 2nd one of the Napoleon castings had been completed and was "in the sand—cooling slowly." By mid-December two Napoleons had been completed and two more were cast. Two additional Napoleons were turned out before the end of the year. By February, 1862, the entire order for fourteen cannon had been completed.[12]

Bragg later ordered Oladowski to procure a sufficient number of guns to complete the batteries then forming in Mobile. Accordingly, an order was placed with the firm on Christmas Day, 1861, for thirteen additional field guns. On December 30th the firm acknowledged the order but stated that they preferred "not to name a definite time for the delivery of these guns, but beg of you to rely on our extreme desire to furnish them at the shortest possible moment." The contract was completed by early April.[13]

Thus, by the end of 1861, the Confederate Government had placed orders with Leeds for a total of twenty-seven field pieces amounting to $20,615.40. All of the guns were completed, accepted, and placed in service by early 1862, yet Government remittances were slow in arriving. At the time the city fell in April, 1862, some $16,610.20 was still outstanding. Full Government payment did not come until a year later. Oladowski, then a

The etching above depicts Southern civilians turning in their bells to be cast into cannon.

(Library of Congress)

Lieutenant Colonel and Chief of Ordnance of the Army of Tennessee, substantiated the firm's claim. He stated (April 1863): "I am certain of their being delivered as I saw them in the Army and some of them are now in the service." [14] Six months later, at the Battle of Missionary Ridge, the Federals reported the capture of six of the Leeds, 12-pounder Napoleons. Three Leeds cannon, two 1861 rifled 6-pounders and a bronze 6-pounder, passed through the Atlanta Arsenal during April and June, 1864.[15]

Leeds & Company also completed a large order for Major M. L. Smith in early 1862. On March 3rd a battery of two 6-pounders and two 12-pounder howitzers was delivered and approved. The next day, March 4th, two six gun batteries complete with carriages, caissons, battery wagons, and forges were delivered. Total cost amounted to $23,672.[16]

Between October, 1861, and April, 1862, Leeds and Company thus made definite contract deliveries of forty-nine bronze field pieces as follows:[17]

DATE OF DELIVERY	DELIVERED	ARSENAL
October 1861	4 6-pounders	New Orleans
	2 6-pounder rifles	
December 3	3 6-pounder rifles	"
	3 6-pounders	
February 19, 1862	2 6-pounder rifles	"
	6 12-pounder Napoleons	
March 3	2 6-pounders	"
	2 12-pounder howitzers	
March 4	8 6-pounder rifles	"
	4 12-pounder howitzers	
April 5	6 12-pounder Napoleons	"
	3 12-pounder howitzers	
	2 6-pounders	
	2 6-pounder rifles	"
	TOTAL CANNON DELIVERED 49	

Casting of bronze ordnance was predicated on the availability of copper and tin and the problem of a raw material deficiency became more acute as production increased. General P. G. T. Beauregard felt that no reliance could be placed upon the Government to relieve the pressing situation. In March, 1862, he, therefore, appealed to the citizens of the Mississippi Valley for their assistance, asking them to donate their plantation and church bells to be melted down for cannon.[18] General Bragg considered this move sensational and unnecessary but the public at large gave its endorsement and response soon came from courthouses, factories, public institutions, plantations, and churches. Wagon loads of andirons, candlesticks, gas fixtures, and even doorknobs were collected and sent to the foundries.[19] The New Orleans *Cresent* went so far as to suggest that if the need for iron continued, that the statues of Andrew Jackson in Jackson Square and Henry Clay in Canal Street be sacrificed for cannon.[20]

Enthusiasm apparently even exceeded practicalities as one Mobile foundryman stated that many of the ornaments and household utensils received were simply worthless. "There is no gun metal in them," stated the mechanic, "and we are daily refusing to accept such articles." [21] Bell metal, on the contrary, proved quite satisfactory and Beauregard was soon negotiating with Leeds & Company for the conversion of the city church bells into 6-pounder rifles and 12-pounder Napoleons.[22]

Many of the bells gathered in were never used. "One of the most striking objects which presented itself as our steamer reached the levee opposite St. Mary's Market, was an immense collection of bells lying on the wharf, covering, it seemed to me, a quarter of an acre, and amounting in numbers to hundreds," reported a Northern correspondent. "These were Beauregard's bells, sent in response to his call. They were all sizes, from very large church bells weighing hundreds of pounds, down to small plantation and steamer bells." The bells were later collected and sent to Boston.[23]

Not only the deficiency but also the rising cost of materials created difficulties. "The dealers in copper (much as we use to make cannon) are still rising their prices," reported Leeds in late 1861. "Lately we have bought old copper at 22¢ and new copper at 26½¢. We have since received notice that the price would be 31½¢ for the next lot. We presume the price of old copper will rise in proportion." These rises were reflected in the prices charged for cannon. In October, 1861, a 6-pounder sold for $625 and a 6-pounder rifle for $675. By March, 1862, the same guns sold for $700 and $800 respectively. The price for 12-pounder howitzers varied from $700–$750 and that of Napoleons from $1125–$1150.[24]

By the spring of 1862 some Leeds cannon were being tested on the battlefield. On April 6, 1862, General Albert Sidney Johnston launched a surprise attack against General U. S. Grant's army encamped at Pittsburg Landing. When the Confederate advance became snagged by unexpected resistance in the Union center, Brigadier General Daniel Ruggles assembled nearly sixty cannon to smash the Union position. Many of these batteries were armed with Southern-made cannon. Bankhead's (Tennessee) and Byrne's (Mississippi) Batteries were armed with Quinby & Robinson guns. Rutledge's (Tennessee) Battery was supplied with six guns from the Brennan foundry in Nashville. The six cannon of the Louisiana Washington Artillery (5th Company) and four Napoleons of Robertson's (Florida) Battery were made by Leeds & Company.[25] The Union position, known as the "Hornet's Nest," fell under this tremendous concentrated artillery fire. The mass capture of 2200 Federals was as much a tribute to Southern industry as it was to Southern artillery and infantrymen.

Even as Shiloh's guns pounded away, however, Leeds & Company was appealing to the military authorities for permission to evacuate New Orleans. A Federal fleet under Admiral David Farragut was threatening the forts on the Lower Mississippi. Permission to remove the machinery was denied, however, for fear that the evacuation would create panic in the city. With the passage of the forts by the Federal fleet, Confederate Commander, Major General Mansfield Lovell, on April 24th, ordered the firm to remove as much work as possible. Some Government

A Leeds 3.3-inch rifle in the West Point Collection. These weapons were referred to in company records simply as rifled 6-pounders. *(West Point Museum)*

work was removed but 200 tons of shot and shell and all the plant's valuable machinery had to be abandoned.[26] J. W. Mallet, Confederate ordnance expert, later remarked that the loss of Leeds was "one of the sorest consequences of the fall of the city." [27]

Leeds & Company remained dormant for the remainder of the war. In 1865 the establishment reopened under the same superintendence, with the exception of Thomas Leeds who died shortly after the capture of the city. A number of the old operatives were rehired and the firm soon regained a prosperous business.[28]

Six Leeds & Company cannon (three 6-pounders and three 6-pounder rifles) are in the West Point collection. One of the rifles and one smoothbore were cast in 1861 and the others in 1862. Although the bronze rifled guns cast by the firm are referred to in company records as 6-pounder rifles, they should perhaps more correctly be termed 3.3-inch rifles. An 1861 Leeds Napoleon, foundry number 18, is at Petersburg, two 1862 Napoleons, numbers 38 and 39, are at the John M. Browning Memorial Museum at Rock Island, Illinois, and a fourth Leeds Napoleon survives in Boone County, Illinois. Two Leeds 6-pounders, one dated 1861 and one 1862, are at Gettysburg. All of the weapons are marked on the right trunnion face: "LEEDS & CO. NEW ORLEANS."

NOTES

[1] New Orleans *Times,* December 30, 1865.

[2] Ibid.; *Gardner's New Orleans Directory for 1861,* 482; *O.R.,* VI, 620.

[3] Dew, *Ironmaker to the Confederacy,* 101.

[4] New Orleans *Picayune,* June 21, 1861.

[5] *O.R.,* VI, 584, 634; *O.N.R.,* Series 2, I, 740.

[6] *Ibid.,* 468; O.R., VI, 621; Charles L. Dufour, *The Night The War Was Lost* (Garden City, N.Y.: Doubleday and Company, Inc., 1960), 64. A Citizen's File voucher indicates that 6 32-pounders were delivered October 16, 1861.

[7] Nashville *Union and American,* May 7, 1861; John D. Winters, *The Civil War in Louisiana* (Baton Rouge: Louisiana State University Press, 1963), 60–62; *O.N.R.,* Series 2, I, 740.

[8] Albaugh and Simmons, *Confederate Arms,* 238.

[9] Leeds & Company, "Citizens File."

[10] *O.R.,* VI, 726; Skates & Company, "Citizens File."

[11] Leeds & Company, Ibid.

[12] Ibid.

[13] Ibid.

[14] Ibid.

[15] *O.R.,* XXXI, pt. 2, 99; Record of Receipts and Issues at Atlanta Armory, December 1863–June, 1864, Chap. IV, Vol. 17, page 335, Vol. 18, page 82, National Archives Record Group 109, War Department's Collection of Confederate Records. (Information courtesy Sydney Kerksis and Edwin Olmstead).

[16] Leeds & Company, "Citizens File."

[17] Ibid.

[18] *O.N.R.,* Series 2, I, 523, 561; Alfred Roman, *The Military Operations of General Beauregard,* 2 Vols. (New York: Harper and Brother, 1884), Vol. I, 225; Mobile *Register,* March 18, 1862.

[19] Bell I. Wiley, *The Life of Johnny Reb: The Common Soldier in the Confederacy* (New York: The Bobbs-Merrill Company, Inc., 1943), 299; Stevenson, *Thirteen Months in the Rebel Army,* 61.

[20] Dufour, *The Night The War Was Lost,* 211

[21] As quoted in Charleston *Mercury,* April 2, 1862.

[22] *O.R.,* X, pt. 2, 362.

[23] Columbus *Enquirer,* July 15, 1862; See also New Orleans *Picayune,* April 9, 1862.

[24] Leeds & Company, "Citizens File."

[25] Ibid.; New Orleans *Picayune,* September 11, 1861.

[26] *O.R.,* VI, 622, 634.

[27] J. W. Mallet, "Work of the Ordnance Bureau," *Southern Historical Society Papers,* XXXVII (1909), 7.

[28] *O.R.,* VI, 621; New Orleans *Times,* December 30, 1865.

The Noble Brothers Foundry, Rome, Ga. *(A History of Rome and Floyd Co.)*

NOBLE BROTHERS & COMPANY

One of the more active cannon foundries in the South was that of Noble Brothers & Company of Rome, Georgia. After the war a Federal officer remarked that "the Noble Brothers . . . did the federal government more harm than any one regiment of rebel soldiers did during the entire war." [1] The activities of the firm, indeed, constitute a fascinating segment of Civil War history.

In 1837 James Noble, an English mining engineer and mineralogist, moved to America and settled in Reading, Pennsylvania. There he established a foundry and machine shop, but the facility was later destroyed by fire. Shortly thereafter he returned to England to see the World's Fair and was attracted to the Southern brand of iron on display. Returning, he toured the South and decided to locate in Rome, Georgia. In the summer of 1855, in partnership with five of his sons—John, James, Samuel, William, and George—he organized and built Noble Brothers & Company. The facility was located near the river at the east end of 3rd Street.[2]

Prior to the war the firm manufactured steam engines and boilers and also constructed a twenty-five ton locomotive, the first such engine built south of Richmond. The foundry later expanded to include iron castings of any size, mining machinery, railroad work, bar iron, grates and pipes. The establishment was the repair shop for the Rome Railroad and the Coosa River Steamboat Company.[3]

At the first signs of hostilities James Noble went to Montgomery, Alabama, to confer with President Davis concerning a Confederate contract. A contract was signed around June, 1861, and the appropriate drawings for 3-inch rifles were furnished.[4] The original contract called for twenty-seven field pieces, although by the end of the summer the number had been raised to fifty-six, mostly 3-inch rifles.[5] An 1861 Noble 3-inch rifle sold for $175, plus $815 for the carriage and caisson.[6] The first battery cast by the Nobles, consisting of three 6-pounders and three 3-inch rifles, went to the Cherokee Artillery, a volunteer unit of Rome.[7]

By the end of August Noble Brothers had shipped a number of field pieces to Richmond. At that time the Rome *Courier* reported:

> MORE RIFLED CANNON—Messrs. Noble Brothers & Company shipped four more of their splendid rifled cannon to Richmond on Saturday last. This makes a full battery they have shipped. They have a number already cast, which they are rapidly finishing up. They are also making a new patent breech loading brass gun which, it is said, will be effective about—well, a greater distance than we believe it will. We, however, believe that the idea is good, and that it will prove an effective weapon.
>
> Since the above was put in type, we learn they shipped four more yesterday, and will ship two more to-day, making three full batteries.[8]

Confederate inspectors were sent to the foundry to inspect and test fire completed weapons. On September

A Noble Brothers 12-pounder howitzer at the Petersburg National Battlefield Park. *(Tarry Beasley)*

10th a test was made on the metal of which the 3-inch rifles were being cast. A gun was loaded with a double service charge of two pounds of powder and a nine pound projectile. Fifteen such rounds were fired. The charge was then increased to three times normal. Three pounds of powder and a fifteen pound projectile were used. Upon the second firing of this severe load the gun burst into three pieces. The metal was thus found quite satisfactory.[9]

Production never kept pace with demand; there were always artillery units waiting to be issued guns. "We are very much in need of three batteries," wrote General D. H. Hill in October, 1861. "I understand they can be purchased in Rome, Georgia, and one battery received fifteen days from the date of order. I most earnestly urge the immediate purchase of these guns." [10]

Early in the war most Noble guns were dispatched to the Army of Northern Virginia. Cutt's (Georgia) Battery, the Jeff Davis (Alabama) Artillery, and Griffin's (Virginia) Light Artillery were all armed with Noble guns.[11] Some guns were later shipped to points across the Confederacy. On October 21st a 3-inch rifle and a 12-pounder howitzer were loaded on the steamer *William Ramsey* for shipment to Calhoun, Georgia. The weapons were destined for Pensacola.[12] A few weeks later the Rome Railroad received four Noble guns for shipment to Bragg at Pensacola and six cannon for General Hardee at Memphis.[13]

By January, 1862, the firm was heavily engaged in the production of cannon and battery equipment. A newspaper at that time reported:

> MANUFACTURE OF CANNON IN GEORGIA.—The Rome *Courier* says:
>
> At this foundery they are now completing on an average, one cannon a day, and in the manufacture of gun carriages, caissons, etc. are fitting up a battery of six guns once every three weeks. With a few other establishments as active and energetic as are the Messrs. Noble, the Confederate States would soon be supplied with abundance of cannon of her own manufacture. We are glad to learn that these guns have been proved to be of the very best character of iron ordnance. The famous Cutt's battery that did such distinguished service at Dranesville, we are told, was from this establishment.[14]

On December 31, 1861, Major L. H. McIntosh placed an order for a complete six gun battery (two 6-pounders, two 3-inch rifles, and two 12-pounder howitzers) for the State of Georgia. Delivery was delayed several weeks due to a lack of iron for the wheels, but by mid-February the entire battery was completed. The guns were rushed to Savannah.[15]

Although strategically located near the rich Alabama coal and iron ore fields and the large Ducktown copper mines near Chattanooga, the Noble foundry was in need of materials at an early date. The difficulty was often due to a lack of proper transportation. "We can get neither cars nor coal," stated the Nobles in April, 1862, "and will soon be obliged to stop." [16] In an effort to increase the supply of raw materials, scrap metal drives were conducted and churches donated their bells for the war effort. "The shopyard was a wonderful sight," reported a member of the Noble family, "with loads of stills and their worms from fifty to hundred gallons, bells from churches, brass and iron fenders, candlesticks and other heirlooms given cheerfully by the farmers." [17]

The shortage of labor also created difficulties. By November, 1861, the firm was advertising for an additional twenty blacksmiths and five machinists.[18] Numerous communications were sent to the Secretary of War requesting additional mechanics but with little effect. In August, 1862, a single molder was ordered to report to the foundry to assist in the manufacture of projectiles.[19]

In February, 1862, Governor Joseph Brown of Georgia exempted twenty men at the plant of military age to protect them from the militia officer.[20] These men, along with other exempted mechanics in Rome, were formed into an artillery company for defensive purposes. A scare came in May, 1862, as the Nobles anxiously wrote Governor Brown that "much and well grounded apprehension is felt for the safety of this place." [21] It was not until May of the next year, however, that the company, armed with two cannon from the foundry, was ordered to the defense of the city.[22] Fortunately, the raiding cavalry, led by Colonel A. D. Streight, was captured by the cavalry of General Nathan Bedford Forrest about twenty miles from the city.

Further complications were experienced when in August, 1862, a fire swept through the gun carriage department. The facility was located in a three-story building also occupied by the Dickson & Nelson Rifle Factory. The fire started in the drying room and spread so rapidly that workmen were forced to jump from the second-story windows. The building and contents, not covered by insurance, were a total loss. The exact cause of the fire was never determined although it was thought to be the work of an incendiary. Gun carriage construction ceased after that date.[23]

Despite these numerous difficulties the Noble Brothers experienced considerable success in the casting of bronze and iron field guns. Between April, 1861, and October, 1862, some fifty-eight field pieces were delivered to the C. S. Government as follows:[24]

DATE OF DELIVERY	DELIVERED	ARSENAL
April 14, 1861	1 6-pounder	Augusta
	1 3-inch rifle	
	2 12-pounder howitzers	
August 27	4 6-pounder rifles	Richmond
	2 12-pounder howitzers	
	4 3inch rifles	
September 3	2 3-inch rifles	"
September 21	1 12-pounder howitzer	"
	1 3-inch rifle	
November 1	2 12-pounder howitzers	"
	2 rifled guns	
November 22	4 3-inch rifles	"
November 23	2 3-inch rifles	"
	2 6-pounders	
	2 12-pounder howitzers	
May 1, 1862	1 3-inch rifle	"
	7 6-pounders	
June 9	3 3-inch rifles	Augusta
	3 6-pounders	
September 20	4 6-pounders	"
	2 12-pounder howitzers	
October 3	3 6-pounders	Atlanta
	3 12-pounder howitzers	
	TOTAL CANNON DELIVERED 58	

The Nobles' production was not limited to field guns. In April, 1862, six 8-inch siege howitzers cast for the State of Georgia were accepted and shipped to Savannah.[25] Only a few weeks later, however, the C. S. Government rejected five similar cannon due to faulty casting.[26] The Nobles protested but the Ordnance Department was unyielding. When Rome was threatened by Streight's Raid in May, 1863, the weapons were buried on the foundry grounds to prevent capture. Both the C. S. Government and the State of Georgia later cancelled their orders with the firm for heavy columbiads.[27]

An unusual incident occurred in October, 1862, which caused a sharp exchange between the Nobles and the C. S. Government. In that month two Ordnance Department representatives, Maynadier Mason and Captain L. J. Smith, inspected the foundry and complained about the faulty casting of several cannon. At that, one of the Nobles began insulting Smith by saying that "all ordnance officers were fools and jackasses." Mason objected and was knocked down and beaten.[28]

The incident, apparently not the first, was promptly reported to Gorgas. The Chief of Ordnance moved swiftly. On November 12th a stern letter was dictated to the Nobles. "In view of your abusive and insulting language against the officers of the Department and your insults to Capt. L. J. Smith, C.S.A., while on discharge of his duties as an inspector of ordnance at your foundry, you are hereby informed that no more of your bills will be paid at this office." [29]

An apology from the Nobles was forthcoming, but the incident was not dropped. Samuel Noble, oldest of the brothers, was well known throughout the region to have strong Union sympathies. In consideration of the overall circumstances, the Government later confiscated some of the machinery at the plant and removed it to Augusta. Casting of cannon ceased after that time.[30]

Many Noble guns were eventually phased out with the advent of more modern weapons. In the Western Theater, however, a number of Noble Cannon were still in use in late 1863. A listing of the guns in Walker's Division, Army of Mississippi, revealed that two Noble 12-pounder howitzers were in use.[31] At the Battle of Chickamauga in September, 1863, Confederate recaptured three Noble cannon—an 1862 12-pounder howitzer and two 3-inch rifles. Seven Nobles were channeled through the Atlanta Arsenal between January and June, 1864. Four of the cannon, one 12-pounder howitzer, one iron 6-pounder and two iron 3-inch rifles, were turned over to the Augusta Arsenal. Two guns, an iron 3-inch rifle and a bronze 6-pounder were issued from the Arsenal. One gun, a bronze 3-inch rifle, was turned in to the Arsenal on July 6th.[32]

During the remainder of the war Noble Brothers was involved only in the production of iron ore in Cherokee County, Alabama. The ore was hauled by wagon to two

An iron 6-pounder cast by Noble Brothers in December, 1861. The weapon is now in private hands.
(Troy L. Church)

Left: JAMES NOBLE, SR.   Right: SAMUEL NOBLE   *(John McCaa)*

furnaces where it was beaten down by Negroes, both men and women, and then melted. The ore was then shipped to the Rome plant via steamboat. Some thirty-four Negroes were employed at the Rome facility in October, 1863.[33]

The Nobles' plant was destroyed when Sherman's troops entered Rome on May 18, 1864. The large smoke stacks of the foundry were blown up with powder blasts and the shops burned.[34] Much of the remaining machinery was saved, however, and on June 1st a twenty-two car train was loaded and sent to Chattanooga. Included in the machinery was one trip hammer, one 12-horsepower engine, one steam engine, one vertical boiler, one screw cutting machine, one horizontal drill press, one slide lathe, one double head slide lathe, one lathe for turning locomotive wheels, one small speed lathe, one circular saw, and 7000 feet of lumber.[35]

After the war the Nobles rebuilt their plant and soon regained a prosperous business. For a number of years they tried unsuccessfully to retain possession of their machinery or to be properly compensated. In an 1873 claims investigation, however, the Nobles' request was rejected. The investigating officer concluded that "when the Federal troops came to Rome 18th of May the Nobles then became very loyal to the Federal Government and was just as ready to contract with it and receive greenbacks as pay as they had been to contract with the rebels and receive its worthless currency." [36]

A fine collection of Noble guns survives. Two 12-pounder howitzers, cast in September and December, 1862, are at Gettysburg. Several bronze 6-pounders survive, including a piece at Kennesaw Mountain. An iron 6-pounder, cast in December, 1861, is in private hands in Winston-Salem, North Carolina. Noble weapons are inscribed on the left trunnion face with the month and year of casting and on the right trunnion face with the firm name: "NOBLE BROTHERS & CO. ROME, GA." Occasionally an "L. J. S." (Captain L. J. Smith) may be inscribed, indicating that the weapon was inspected by that officer.

NOTES

[1] Report of E. Richmond, September 1873, House Files of the Southern Claims Commission, National Archives. Hereafter cited as "House Files."

[2] Ethel Armes, *The Story of Iron and Coal in Alabama* (Birmingham: Chamber of Commerce, 1910), 184–185.

[3] Albaugh and Simmons, *Confederate Arms,* 250; Rome *News-Tribune,* July 10, 1964.

[4] Kate Quintard Noble Roberts, "A War Time Foundry," *The Alabama Historical Quarterly,* XVIII (1956), 463–464.

[5] Charleston *Courier,* July 3, 1861; Lee, ed., *The Memoirs of William Nelson Pendleton,* 168.

[6] Noble Brothers & Company, "Citizens File."

[7] Rome *Courier Weekly,* July 13, 1861.

[8] As quoted in Columbus *Inquirer,* August 31, 1861.

[9] Rome *Courier Tri-Weekly,* September 12, 1861.

[10] *O.R.,* IV, 677.

[11] "House Files"; Richmond *Dispatch,* September 16, 1861; Noble Brothers & Company, "Citizens File."

[12] Ibid.

[13] Ibid.

[14] As quoted in Memphis *Appeal,* January 24, 1861.

[15] McIntosh to Noble, December 31, 1861, State Ordnance Department to Noble, January 8, 1862, Waters to Noble, January 29, 1862, Noble to Brown, January 29, 1862, Brown to Noble, February 3, 1862, Georgia State Archives.

[16] Noble to Brown, April 14, 1862, Ibid.

[17] Roberts, "A War Time Foundry," 463.

[18] Columbus *Weekly Sun,* November 12, 1861.

[19] Noble Brothers & Company, "Citizens File."

[20] Allen D. Chandler, ed., *The Confederate Records of the State of Georgia,* Vol. II (Atlanta: Charles P. Byrd, 1909), 201–202.

[21] Noble to Brown, May 20, 1862, Georgia State Archives.

[22] Noble Brothers & Company, "Citizens File."

[23] Rome *Courier Weekly,* August 31, 1862.

[24] Noble Brothers & Company, "Citizens File."

[25] Noble to Brown, April 14, 1862, Georgia State Archives.

[26] Noble Brothers & Company, "Citizens File."

[27] McIntosh to Noble, June 1, 1862, Georgia State Archives; *O.R.,* XIV, 604.

[28] Noble Brothers & Company, "Citizens File."

[29] Albaugh and Simmons, *Confederate Arms,* 250.

[30] Noble Brothers & Company, "Citizens File"; Samuel Noble to Thomas, September 2, 1865, Union Provost Marshal's "Citizens File" pertaining to Samuel Noble, National Archives; Rome *Courier Weekly,* February 18, 1863.

[31] Listing of Cannon in Walker's Division, Army of Mississippi, 1863 (month unknown), Mississippi State Archives.

[32] *O.R.,* XXX, pt. 2, 41–42; Record of Receipts and Issues at the Atlanta Armory, December, 1863–June, 1864, Chap. IV, Vol. 17, pp. 139, 314. Vol. 18, pp. 82, 118, 164, National Ordnance Records Group 109, War Department's Collection of Confederate Records. (Information courtesy of Sydney Kerksis and Edwin Olmstead.)

[33] Roberts, "A War Time Foundry," 464; Noble Brothers & Company, "Citizens File."

[34] George M. Batty, Jr., *A History of Rome and Floyd County* (Atlanta: The Webb & Vary Company, 1922), 177–178.

[35] "House Files."

[36] Report of E. Richmond, September 1873, Ibid.

WM. T. QUINBY. J. L. CLAGHORN. W. A. ROBINSON.

WESTERN FOUNDRY,

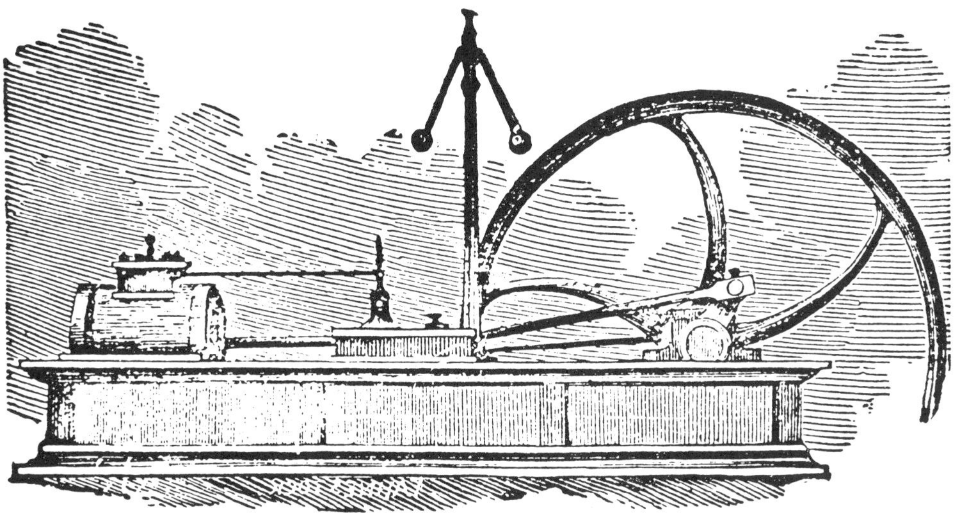

Poplar Street, front of Exchange Building,

MEMPHIS, TENN.

QUINBY & ROBINSON,

BUILDERS OF

STEAM ENGINES, SAW MILLS,

Gin and Mill Gearing, Plows, &c.

COPPER, BRASS AND SHEET IRON WORK DONE.

Plain and Ornamental Castings, for Buildings,

MADE TO ORDER.

Having the latest and most approved Patterns for COLUMNS, WINDOW CAPS, TRUSSES, BRACKETS, and other Ornamental Work. STEAMBOAT, RAILROAD, and other Repairing, done with dispatch. Orders solicited and will be promptly attended to.

An 1860 Quinby & Robinson Advertisement *(Memphis City Directory)*

QUINBY & ROBINSON

A careful observer at the Shiloh National Military Park will notice that five field pieces are inscribed: "QUINBY & ROBINSON MEMPHIS, TENN." Quinby & Robinson was one of the most notable cannon foundries in the Confederacy, a fact known only to the historian far from the beaten path.

Quinby & Robinson was under the ownership of William T. Quinby and William A. Robinson. The firm, located on Front Street at the foot of Poplar, was the successor to Curtis & Knapp and A. G. Knapp & Company, a long established Memphis firm specializing in steam engines, saw mills, gin and mill gearing, and a variety of copper, brass, sheet and iron work.[1] Quinby and Robinson assumed proprietorship in 1859, and by 1860 the white labor force included forty-one hands—one bookkeeper, one core maker, one engineer, seven finishers, two blacksmiths, thirteen molders, and twelve laborers.[2]

By April, 1861, a portion of the hands were busily engaged in the casting of a 32-pounder seacoast gun.[3] Difficulties were encountered during the casting process, however. As the liquid metal was poured into the mold, the sand gave way, choking up the entrance of the mold.[4] The project was apparently scrapped.

The first order for field guns came from Mississippi, when Byrne's independent company of artillery at Greenville contracted for a battery of six bronze pieces—two 12-pounder howitzers and four 6-pounders.[5] By May 31st one of the 6-pounders had been cast and was nearing completion.[6] On the 4th of June the gun was taken to the upper levee for testing. The trial proved a success, although much of the populace was frightened, supposing the shots to be those of the city alarm guns signaling the approach of the enemy.[7] By June 14th the firm had received an order from General Gideon Pillow for two 6-pounders for Tennessee.[8]

Production of field pieces soon began on a larger scale. In August a correspondent from the Memphis *Appeal* visited the plant and reported the following:

> WARLIKE—We looked into Quinby & Robinson's shop yesterday and found eight beautiful brass cannon ready for the carriage, and twelve more in various states of fowardness. There were also three or four iron cannon on hand, one of them of the peculiar kind known as the Parrott gun, a long slendor piece which throws a shot of ten inch in length and two and three quarters in diameter. The same manufactory are making a boring machine for the Confederate States arsenal by which twelve rifles will be bored at once; they are also putting up machinery for rifling cannon.[9]

Many of these guns were excellent castings. In September, 1861, a shipment of Tredegar 6-pounders arrived at the Memphis levee. A Memphis reporter proudly stated: "For beauty and finish and general appearance the Richmond cannon cannot compare with those made here and we have heard military men say that Messrs. Quinby & Robinson's guns cannot be excelled by any in the world." [10]

The production of iron and bronze cannon required large stores of raw materials. On June 24, 1861, an installment of 19,000 pounds of copper was received from the Hope County Mine in Tennessee.[11] On October 29th the proprietors requisitioned the Ordnance Department in Memphis for additional copper, but it was November 21st before a shipment of 10,016 pounds was received. On January 28, 1862, a requisition for 700 pounds of block tin, 7000 pounds of copper, and 300 pounds of zinc was immediately filled.[12]

Inadequate transportation often made it difficult to ship materials even when they were available. On December 14, 1861, Quinby and Robinson wrote General Leonidas Polk that a supply of pig iron could be purchased at the Cumberland and Tennessee Rivers, but railroad officials would not spare the cars unless ordered.[13]

By September, 1861, the foundry was turning out several cannon each week.[14] In that month a workman was sent to Columbus, Kentucky, and performed repair work on several batteries. For his services the company was paid $75.00.[15] During the fall months the firm also continued its production of pre-war products. Advertisements were placed as far away as New Orleans announcing the sale of cylinders, boilers, saw mills, grist mills, and all kinds of machinery manufactured "from the best quality of Tennessee and Missouri pig iron." [16]

On the night of September 30th a disaster struck the plant. At 8:00 a fire broke out in the engine room located in the basement of the building. Several of the city fire engines promptly responded to the alarm but a supply of water was not available. The fire quickly spread to a number of wooden patterns and a large supply of coal stored in the basement. From there the flames leaped to the machine shop on the floor above and finally to the pattern makers shop on the third floor. The windows of the foundry soon caught fire endangering that structure. The walls of the foundry eventually collapsed which prevented the fire from spreading further. A steam engine finally succeeded in setting up a hose at a pond some distance away, and the steady stream of water saved the other buildings.[17]

The cause of the fire was never determined. Some thought it to have been started by incendiaries and others by the carelessness of an engineer. Damage was estimated at $60,000. The machine shop contained twelve lathes, one planning machine, and three drill presses and the pattern shop housed patterns, tools, two circular saws, and a gear cutting machine, the only good one in the city. Also destroyed were three engines and a stock of 3000 barrels of coal. Six completed 6-pounders were rescued and several unfinished ones were not damaged, not having been bored.[18]

Arrangements had just been completed to fill a heavy contract with the Confederate Government for cannon, shot and shell and this necessitated a delay. Rebuilding was immediately commenced while Quinby and Robinson went to Charleston and Savannah to procure machinery replacements. In October the owners wrote General Polk requesting an advance on their contract to the amount of $25,000, to be repaid in work, bonds, or securities.[19]

Difficulties were many but the plant was soon put back in a state of production. The machine shop was enlarged in the process by one-half, which enabled the increased production of cannon. During the construction interim some cannon were sent to A. B. Reading & Brother and A. M. Paxton & Company, both of Vicksburg, for finishing work. With construction problems overcome the plant was soon at work for the Confederate Government.[20]

Between November, 1861, and June, 1862, definite contract deliveries to the Confederate Government were made as follows:[21]

DATE OF DELIVERY	DELIVERED	ARSENAL
November 19, 1861	7 6-pounders	Memphis
December 6	2 12-pounder howitzers	"
December 7	2 3-inch bronze rifles	"
December 13	1 12-pounder howitzer	"
	1 6-pounder	
December 18	3 12-pounder howitzers	"
	3 6-pounders	
January 10, 1862	1 3-inch bronze rifle	"
January 10	1 6-pounder (iron)	"
January 15	1 12-pounder Siege Gun	"
January 21	3 12-pounder howitzers	"
January 21	1 12-pounder Siege Gun	"
February 2	2 12-pounder howitzers	"
February 6	1 3-inch bronze rifle	"
February 8	2 12-pounder howitzers	Columbus
February 10	2 12-pounder Napoleons	Memphis
	2 24-pounder howitzers	
February 11	2 12-pounder howitzers	Columbus
February 14	1 12-pounder Siege Gun	Memphis
February 17	3 12-pounder howitzers	Columbus
February 18	1 12-pounder Siege Gun	Memphis
February 20	1 12-pounder Siege Gun	"
February 24	3 12-pounder howitzers	Columbus
February 28	3 12-pounder howitzers	"
March 6	3 12-pounder howitzers	"
March 10	3 12-pounder howitzers	"
March 14	2 12-pounder howitzers	"
June 11	3 24-pounder iron guns	"
	2 12-pounder howitzers	
	2 12-pounder Napoleons	
	1 12-pounder iron gun	
June 11	4 12-pounder howitzers	"
	4 12-pounder Napoleons	
June 12	3 12-pounder howitzers (unfinished)	"
	1 6-pounder (unfinished)	
	TOTAL CANNON DELIVERED 77	

An 1862 Quinby & Robinson 12-pounder howitzer at the Shiloh National Military Park. *(Tarry Beasley)*

By the end of 1861 the Confederate Government had paid the firm over $32,000 for work delivered—$17,000 in cash and $15,000 in copper and material equivalents. The average cost of a 6-pounder was $685.10 and that of a 12-pounder howitzer was $610.70.[22]

In January, 1862, Quinby and Robinson received an unsigned letter warning the owners of an association of spies sent to destroy their foundry. The spies, one of whom had been with John Brown on the infamous Harper's Ferry Raid, had reportedly been in the foundry and knew their way around. "Rely on it," warned the letter. "This information is positive and undoubted and I trust will not only be in time to save you but that you will succeed in catching and causing the hanging or burning these devils at the stake. . . ." [23]

This was possibly part of a much larger association of spies organized by Federal authorities and sent South to destroy the primary foundries. The plot was discovered by an anonymous Louisville clerk and later capitalized upon by Southern newspapers.[24] Whatever developed from this group is not known, but there are no records of any foundries in the Western Confederacy having been sabotaged.

After the fall of Fort Donelson on February 16, 1862, Confederate authorities became alarmed over the safety of Tennessee and began making arrangements to have all ordnance property removed to safer places. As early as mid-February Quinby and Robinson began delivering cannon to Columbus, Mississippi. The proprietors also requested transportation of their machinery by steamboat to Grenada or Vicksburg and thence by rail to Atlanta. In a note to General Beauregard, General Leonidas Polk wrote: "Messrs. Quinby and Robinson are the proprietors of one of the largest foundries in the South West and it is of the utmost importance to the country that their machinery etc. should be saved." [25]

Just prior to the fall of Memphis to the Federal Navy on June 6, 1862, Quinby and Robinson sent a large shipment of war material by rail to Columbus. The equipment included some machinery (patterns, gun flasks, four lathes for boring cannon, etc.), over 100,000 pounds of coal, 416 bushels of coke, hundreds of rounds of shot and shell, and twenty cannon, four of which were unfinished castings.[26]

After the evacuation of the Memphis plant Quinby and Robinson bought out the Etowah Iron Works near Cartersville, Georgia, from Mark A. Cooper. The latter part of 1864 found Quinby at his southwest Georgia plantation in ill health.[27] After the war the firm, under the leadership of William Robinson, continued peacetime production at the corner of Hotel and Shelby Streets in Memphis. Products included "all kinds of machinery and castings, house fronts, steam engines, and everything made of iron and brass." [28]

Twelve Quinby & Robinson field pieces survive. Two 1862 12-pounder Napoleons, numbers 37 and 38, and

An 1862 Quinby & Robinson bronze 3-inch rifle surviving at Petersburg National Battlefield Park. The firm cast only four of these weapons, two of which survive. *(Tarry Beasley)*

two 1861 12-pounder howitzers with no foundry numbers are at Gettysburg. Two 1862 bronze 3-inch rifles are at Petersburg. At the Shiloh park are two 1862 24-pounder howitzers, one of which is number 35. One of these weapons has "A. W. McCONNELL" inscribed on the right trunnion. That name appears in some of the company records and the 1857 Memphis Directory lists him as bookkeeper of the firm. He was possibly an ordnance inspector at the time. Also at Shiloh are one 1861 and two 1862 12 pounder howitzers, numbers 18, 32, and 41 respectively. An 1862 12-pounder howitzer with no foundry number is in private hands in New Jersey. All of the cannon are stamped on the right trunnion face with the firm name: "QUINBY & ROBINSON MEMPHIS, TENN."

NOTES

[1] *Twyman's Memphis Directory and General Business Advertiser for 1850,* 15, 25.

[2] *Williams' Memphis City Directory, City Guide, and Business Mirror,* 56–337.

[3] Memphis *Appeal,* April 20, 1861.

[4] *Ibid.,* April 24, 1861.

[5] Memphis *Avalanche,* May 31, 1861; Ed Porter Thompson, *History of the Orphan Brigade* (Louisville: Lewis N. Thompson, 1898), 857.

[6] Memphis *Appeal,* May 31, 1861.

[7] *Ibid.,* June 5, 1861.

[8] *Ibid.,* June 14, 1861.

[9] Memphis *Appeal,* August 20, 1861.

[10] Memphis *Avalanche,* September 21, 1861.

[11] Memphis *Appeal,* June 25, 1861.

[12] Quinby & Robinson, "Citizens File."

[13] Ibid.

[14] Memphis *Appeal,* October 1, 1861.

[15] Quinby & Robinson, "Citizens File."

[16] New Orleans *Picayune,* September 10, 1861.

[17] Memphis *Appeal,* October 1, 1861.

[18] *Ibid.;* Memphis *Avalanche,* October 1, 1861.

[19] Quinby & Robinson, "Citizens File."

[20] Ibid.

[21] Ibid.

[22] Ibid.

[23] Ibid.

[24] Vicksburg *Whig,* February 25, 1862; New Orleans *Picayune,* March 2, 1862.

[25] Quinby & Robinson, "Citizens File."

[26] Ibid.

[27] Ibid.; Cooper to Ames, August 25, 1862, Secretary of War Letters Received, War Department Collection of Confederate Records.

[28] *Halpin's Memphis City Directory, 1866,* 57, 191.

A. B. READING & BROTHER

In March, 1862, General P. G. T. Beauregard wrote Chief of Ordnance Josiah Gorgas concerning the chain of cannon foundries developing throughout the Mississippi Valley. He stated: "I regard it as clearly advantageous to encourage the casting of such guns at different points in this valley, so that should a foundery unfortunately fall into the hands of the enemy we should not be wholly crippled and deprived of our resources but have several centers of manufacture." [1] Thus, while many of the foundries that had developed in the initial stages of the war slowed due to insufficient raw materials needed to feed them, additional foundries were converted to cannon production. Such was the case with A. B. Reading & Brother of Vicksburg, Mississippi.

Abram Breech Reading was a large Vicksburg landowner as early as the 1830's. Besides the foundry, he owned a large plantation, a brick making plant, and half interest in a saw mill. Little is known of C. A. Reading, Abram's brother. The foundry, located on the northwest corner of Washington and First Streets, turned out steam gauges, circular saw mills, water and gas pipes, and boilers.[2] Reading & Brother later amalgamated with A. M. Paxton & Company, another Vicksburg foundry located at the corner of South and Pearl Streets. During 1861–62, however, these two foundries were separate operations.

By late 1861 the foundry had converted to cannon production and operation was underway. An optimistic Vicksburg *Whig* reported in late December that the facility was geared to produce smoothbore and rifled cannon at the rate of eight to ten per week.[3] Actual production fell far below that level.

In January, 1862, the New Orleans *Picayune* ran a brief article concerning the foundry:

> CANNON FOUNDRY IN VICKSBURG—The Vicksburg *Sun* says that Messrs. A. B. Reading & Brother, proprietors of the foundry in that city that bears their name, are perfecting arrangements which will enable them to cast cannon and ball equal to every respect to those cast at the best foundries in the United States. They have tendered this branch of their foundry to Governor Pettus, giving him carte blanche to use it in any manner that may be deemed essential to the good of the State and the interest of the South.[4]

Local residents expressed a great interest in their new industry. "The children had told me . . . that Father was going to take them to the foundry, to see cannon and bombs cast . . . ," wrote Emma Crutcher to her brother William on Christmas Day, 1861. "Reading's foundry has an order for sixty brass cannon and 30,000 bombshells . . . We saw cannon in all the various stages of completion and I was very much interested. Several other parties of ladies were there." [5]

It was not long before Union intelligence reports on Vicksburg began mentioning that "there is a large iron foundry—casting guns and shot—on the banks of the river in the lower part of town, carried on by one A. B. Reading, probably a traitor by force of circumstances." [6]

Reading began deliveries of cannon to the Confederate Government in December, 1861. Between that time and April, 1862, definite contract deliveries were made as follows: [7]

An A.B. Reading & Bro. bronze 3-inch rifle at Petersburg. *(Tarry Beasley)*

DATE OF DELIVERY	DELIVERED	ARSENAL
December 14, 1861	1 6-pounder	Memphis
December 31	1 6-pounder	"
	1 6-pounder rifle	
January 6, 1862	3 6-pounders	"
	3 6-pounder rifles	
January 11	5 6-pounders	"
	1 12-pounder howitzer	
January 23	4 6-pounders	"
January 30	3 6-pounders	"
January 31	2 rifled bronze guns	"
	1 12-pounder howitzer	
February 6	2 rifled bronze guns	"
February 10	2 rifled bronze guns	"
February 13	1 rifled bronze gun	"
March 25	3 6-pounders	Columbus
	3 3-inch bronze rifles	
April 12	6 6-pounders	"
May 20	3 6-pounders (unfinished)	"
	TOTAL CANNON DELIVERED 45	

All of Reading's castings were of bronze. The 6-pounders produced by the firm weighed 884 pounds, standard weight of the U.S. Model 1841 6-pounder. The 6-pounder rifles were somewhat heavier than standard, weighing in variously at 957, 966, 959, and 935 pounds. The bronze 3-inch rifles were extremely light at 775 pounds apiece. Reading's foundry also cast a number of artillery projectiles and supplied some machinery for the Atlanta Arsenal.[8]

In the spring of 1862 General P. G. T. Beauregard issued an appeal to citizens to turn in their bells to be melted down for cannon. The Vicksburg *Whig* reported that "there are bells enough on the Mississippi between Memphis and New Orleans, and contiguous, to put in the field 30,000 men. Our foundries will receive the bells free of charge, and our steamboats and railroads will no doubt transport them free." [9] Some bell metal was used by Reading. An April, 1862, voucher stated: "C.S.A. to A. B. Reading & Brother for casting and finishing 6 6-pounder bronze guns as per contract with Colonel W. T. Withers, he furnishing 1100 pounds of bell metal to each gun at $350 apiece—$2100." [10]

Some specific Reading weapons are traceable today through records and surviving weapons as follows:

TYPE	NUMBER	REMARKS
3-in. bronze rifle	3	Survives at Petersburg
6-pounder	9	Issued from Atlanta Arsenal on June 15, 1864
6-pounder	10	Survives in private hands
6-pounder	11	Survives in private hands
6-pounder	14	Shipped through Atlanta Arsenal to Augusta Arsenal on April 19, 1864
6-pounder	16	Shipped through Atlanta Arsenal to Etowah, Ga., on Aug. 23, 1863
6-pounder rifle	26	Survives at Petersburg
3-in. bronze rifle	33	In Walker's Div., Army of Mississippi, as of late 1863
6-pounder	35	Survives at West Point
12-pounder howitzer	—	Reported recaptured at Chickamauga in Sept. 1863

An 1862 Reading 6-pounder in the West Point collection.

(West Point Museum)

Surviving weapons are inscribed on both trunnion faces, for example: "A.B.R. AND BRO. V.BURG NO. 3 1862." Unfortunately, the "AND" was stamped so close to the "R" on some of the guns that many ordnance buffs were led on a chase for a mythical "RAND" foundry. All surviving weapons are marked 1862.

NOTES

[1] *O.R.*, X, pt. 2, 362.

[2] Vicksburg *Evening Post,* April 13, 1959; Vicksburg *Daily Whig,* July 1, 1862.

[3] As quoted in Memphis *Appeal,* January 5, 1862.

[4] New Orleans *Picayune,* January 19, 1862.

[5] Emma Crutcher to William O. Crutcher, December 25, 1861, Phillip Crutcher Collection, Mississippi State Archives.

[6] *O.N.R.*, Series 1, XXIII, 227.

[7] A. B. Reading, "Citizens File."

[8] Ibid.

[9] As quoted in Memphis *Appeal,* March 20, 1862.

[10] A. B. Reading, "Citizens File."

[11] Record of Receipts and Issues at the Atlanta Armory, December 1863–June 1864, Vols. 17–18, National Archives Record Group 109 (War Department's Collection of Confederate Records). Information courtesy of Edwin Olmstead and Sydney Kerksis.

[12] Listing of Cannon in Walker's Division, Army of Mississippi, 1863 (month unknown), Mississippi State Archives.

[13] *O.R.*, XXX, pt. 2, 41.

STREET, HUNGERFORD & COMPANY

In Memphis, Tennessee, the foundry of Street, Hungerford & Company, located at the south end of the Navy Yard, converted to cannon and munitions production at an early date. The proprietors, Anthony S. Street and Fayette H. Hungerford, employed nearly 100 hands in the production of wagons, railroad cars, plows, and iron castings.[1]

Street, Hungerford & Company produced a wide variety of ordnance. As early as January, 1861, the owners sent telegrams to various Southern governors indicating a desire to cast shot and shell.[2] Activities were enlarged after the firing on Fort Sumter and by mid-July a dozen iron 6-pounders had been cast.[3] By October the firm was engaged in the construction of ten Hughes' guns, a small breechloader firing a one pound ball six to eight times a minute. A battery of these weapons was used in and around New Madrid, Missouri, and further south in Arkansas.[4] Three Parrott guns were also delivered in October, and it is possible that others were cast.[5]

The quality of some of these early products was so poor as to invoke the scorn of experts. "Many of these guns were defective and even dangerous," wrote one ordnance officer. "One battery from the Memphis foundry [Street, Hungerford & Company] lost three guns in a month by bursting, one of them at the Battle of Belmont, November 7." [6]

Records indicate that a few heavy guns were also cast. On February 18, 1862, "3 guns weighing 30 tons" were delivered, although specific calibers were not mentioned. In May two heavy guns were delivered—a 24-pounder and a mortar "not quite finished." The Ordnance Department was billed $1900 apiece for the weapons.[7] An ordnance officer later recalled that "Memphis supplied some 32 and 64-pounders, also a number of Parrott guns. These were cast at the navy yard firm of Street & Hungerford. . . ." [8] Old heavy ordnance was also rifled and repaired in the shops.[9]

The firm produced a wide variety of wood products prior to the war and therefore made an easy transition to the manufacture of gun carriages. By September, 1861, the plant had completed nineteen gun carriages, and many were turned out after that date.[10] Some field carriages were made for the guns cast at the nearby Quinby & Robinson plant.[11]

There are no known surviving cannon.

NOTES

[1] *Williams' Memphis City Directory, City Guide, and Business Mirror, 1860,* 50–340.

[2] Street, Hungerford & Company to Pettus, January 14, 1861, Pettus Papers, Mississippi State Archives.

[3] Memphis *Appeal,* July 20, 1861.

[4] *Ibid.,* October 10, 1861; *The Confederate Veteran,* XVI (1908), 44.

[5] Street, Hungerford & Company, "Citizens File."

[6] Stevenson, *Thirteen Months in the Rebel Army,* 61.

[7] Street, Hungerford & Company, "Citizens File."

[8] Stevenson, *Thirteen Months in the Rebel Army,* 61.

[9] New Orleans *Bee,* October 7, 1861.

[10] Street, Hungerford & Company, "Citizens File."

[11] Thompson, *History of the Orphan Brigade,* 857.

WASHINGTON FOUNDRY

William J. Hubard—British-born, sculptor, artist, and scientist—is remembered best for his burning desire to copy Houdon's statue of George Washington so that it could be appreciated outside of Richmond. To that end he obtained exclusive duplication rights for seven years by an 1853 act of the Virginia Legislature. Hubard found no foundry capable of producing a suitable bronze casting so in 1855 he founded his own facility—appropriately called the Washington Foundry. It was located just outside of Richmond, Virginia.[1]

With the advent of war Hubard expressed a desire to cast bronze cannon for the State of Virginia. "I will forthwith make two sample guns which shall be put to the most severe test," he wrote Captain Dimmock in April, 1861. "You know that I am better prepared by practical experience for that class of arm (brass) and am willing to take any risk. . . Think on this seriously and let the Washington Foundry have an opportunity of turning out guns to defend the State." [12] A few field pieces were later cast under a State contract with Virginia.

In October, 1861, Chief of Ordnance Josiah Gorgas granted the firm a Confederate contract to cast ten bronze 12-pounder howitzers at a cost of 60¢ per pound. To facilitate completion of the contract an advance loan was

A Washington Foundry 12-pounder howitzer surviving at Shiloh National Military Park. *(Tarry Beasley)*

granted for the installment of additional flasks and lathes.[3]

Hubard's foundry was equipped only to make the rough castings. The weapons were sent to the Richmond machine shop of Thomas Samson and James Pae to be bored, turned and finished. The Samson & Pae firm charged $130.00 apiece for finishing work on the howitzers. The weapons were completed in February, 1862.[4] It was in that month that fate struck the talented Hubard. While experimenting with powder he was killed by an accidental explosion.[5]

At least two Hubard weapons are known to survive. An 1861 12-pounder howitzer is at Gettysburg. It is inscribed on the right trunnion face: "W.J.H. (William J. Hubard) W.F. (Washington Foundry) S & P (Samson & Pae). Also inscribed is "STATE OF VA.", indicating that the weapon was made under State contract. A 12-pounder howitzer with identical markings was issued from Atlanta Arsenal on March 14, 1864.[6] An 1862 12-pounder howitzer, foundry number 3, is at Shiloh and is apparently one of the howitzers made under the Confederate contract. It is marked on the right trunnion face: "W.J.H. & S & P".[7]

NOTES

[1] James C. Hazlett, "Artillery in the Field," Gettysburg Tour: The Lincoln-Civil War Society of Philadelphia (June 1968), 8.

[2] W. J. Hubard, "Citizens File."

[3] Ibid.

[4] Samson & Pae, Ibid.

[5] Hazlett, "Artillery in the Field," 8.

[6] Record of Receipts and Issues at Atlanta Armory, December, 1863–June, 1864, Chap. IV, Vol. 17, pp. 276, National Archives Records Group 109, War Department's Collection of Confederate Records. (Information courtesy of Sydney Kerksis and Edwin Olmstead.)

[7] For years the "W.J.H. & S & P" marking mystified ordnance buffs. The mystery was finally solved in 1965 by the mutual work of Dr. James Hazlett and Mr. Warren Ripley.

S. WOLFF & COMPANY

In the fall of 1861 Samuel Wolff, a wealthy New Orleans merchant and importer, approached Major General Mansfield Lovell with the prospect of investing a large sum of money in a foundry operation. With Lovell's approval, Wolff purchased a foundry, enlarged and completed the facility, bought additional machinery, and employed over 100 hands. The total investment was nearly $100,000.[1]

In November the patterns for 10-inch mortars and field pieces were begun. Patterns for 10-inch columbiads were completed about the time the city fell. The Government furnished the firm with a large lathe for the turning of heavy guns. A reverberating furnace was in course of construction when the city fell. Wolff, who was not acquainted with the casting of cannon, relied heavily upon the knowledge of his employees and local artillery officers. Lovell frequently offered financial assistance to Wolff but the proprietor always declined.[2]

The only known Wolff & Company survivor—an 1862 12-pounder howitzer at the Shiloh National Military Park. *(Tarry Beasley)*

On March 5, 1862, General Lovell took charge of the plant on behalf of the C. S. Government. This was done at the request of Wolff in order to protect his employees from local militia officers. The foundry had done work exclusively for the Government from the beginning.[3]

During the war Wolff testified that "we cast quite a number of field pieces (probably six were finished) and several 10-inch mortars, one of which was completed." Two 10-inch mortars were cast by the firm, both of which were hurriedly thrown into a canal basin upon the evacuation of the city on April 24, 1862. Wolff further stated that at the evacuation "there were, I think, about eight brass pieces (part finished and part unfinished) sent up the Jackson Railroad." Company records indicate that only six field pieces were evacuated—one finished and two unfinished 6-pounders and one finished and two unfinished 12-pounder howitzers. Records give indication of only one field piece having been shipped prior to the evacuation—a bronze rifled gun caliber unspecified.[4]

The enemy took possession of the plant's machinery, tools, a large stock of wood for patterns, and a large stockpile of iron. General Lovell later commented that "in a short time they [Wolff & Company] would have been in condition to contribute very materially to our resources for defense." Time was always a factor to be reckoned with for the South, however, and the idea for cannon production simply came too late to be of any real value. After the fall of the city Wolff moved to Columbus, Georgia. He returned to his importing business in New Orleans after the war.[5]

A lone Wolff survivor, a bronze 12-pounder howitzer, is at the Shiloh park. The weapon is inscribed on both trunnion faces: "S. WOLFF & CO. N.O. 1862."

NOTES

[1] S. Wolff & Company, "Citizens File"; *O.R.*, VI, 606.
[2] *Ibid.*
[3] *Ibid.*
[4] *Ibid.*
[5] *Ibid.*

J. R. YOUNG & COMPANY

In the early months of the Civil War dozens of private foundries across the Confederacy converted to a war economy. "Indeed, every city of the South having a foundry of any size, boasts of furnishing some cannon," stated one officer.[1] While the individual production of many of these foundries was comparatively small, corporately their production was significant.

J. R. Young & Company of the Madison Iron Foundry at Huntsville, Alabama, was organized in 1855. The foundry was located on Mill Street, just behind the Memphis & Charleston Railroad Depot, and turned out mill machinery, stoves, and gin gearings.[2]

Casting of cannon commenced in the early summer of 1861. "A few days ago we were shown by Mr. James Crawford at the Huntsville Machine Shop of the Memphis and Charleston Railroad three beautiful iron 6-pounders which were cast in our city in the foundry of our enterprising fellow citizen J. R. Young & Company and bored out and finished at the machine shop under Mr. Crawford's directions," reported the Huntsville *Democrat* in July. Gun carriages for the weapons were also made at the machine shop.[3] The railroad shop continued to do finishing work on guns cast at the Young foundry throughout the summer. By mid-August seven iron 6-pounders had been completed and mounted. Young was preparing rifling machinery at that time.[4]

In November Dr. John B. Read wrote Brigadier General Withers at his Mobile headquarters that "they [Wright & Rice of Florence, Alabama] use a blast furnace, but so does the foundry at Huntsville from which Col. Gorgas had ordered a number of guns for Richmond." [5] Confederate officials often overestimated the actual production capability of private facilities and the Huntsville foundry proved no exception. In November Samuel Tate, aide to General Albert Sidney Johnston, reported that "I can get six guns a week cut and bored at Huntsville, from 6-pounder to 24-pounder howitzer." [6]

On February 8, 1862, A. J. Hopper, an official of the Memphis & Charleston Railroad, informed General Johnston that the Company's shop at Huntsville had on hand "one 6-pounder Parrott rifled and carriage but not wheels; also, one not mounted at all. . . ." [7] It is very possible that these guns were cast by Young and were at the machine shop awaiting finishing and mounting. Huntsville fell to Union troops on April 11, 1862.

The foundry, although still in existence in the early 1870's, eventually gave way to a cotton mill which was burned in 1881. The Memphis & Charleston Railroad was sold in 1898 and thereafter became a part of the Southern Railroad.

NOTES

1 Stevenson, *Thirteen Months in the Rebel Army,* 62.
2 Huntsville *Times,* April 29, 1974.
3 Huntsville *Democrat,* July 17, 1861.
4 Columbus *Weekly Sun,* August 20, 1861.
5 Wright & Rice, "Citizens File."
6 *O.R.,* IV, 152.
7 *Ibid.,* VII, 156.

THE GOVERNMENT FOUNDRIES

AUGUSTA ARSENAL

The modernization of the field artillery in the Western armies progressed much more slowly than in the East. In January, 1863, the Army of Mississippi, stationed at Grenada, possessed fifty-six field pieces, only two of which were 12-pounder Napoleons.[1] There was a noticeable shortage of Napoleons and Parrott rifles throughout the Vicksburg Campaign.[2] The situation was scarcely better in the Army of Tennessee. The smoke had barely cleared from the Murfreesboro battlefield when battery commanders began complaining of their inadequate armament.[3] In March, 1863, the Army of Tennessee possessed 125 cannon, of which only sixteen were light 12-pounders.[4] The majority of the field batteries still consisted of 6-pounders and 12-pounder howitzers.[5]

In March Bragg ordered a changeover in his artillery. "The commanding general of the army has instructed me to procure 12-pounder light guns and exchange them for 6-pounders and 12-pounder howitzers now in service," reported H. Oladowski, the Army's Chief of Ordnance. "I applied to Colonel Gorgas for them. Two batteries of such guns are ordered to be sent to this army. The balance will be fowarded as soon as they can be procured." [6]

Since nearly all private cannon foundries in the West had been placed out of commission by the end of 1862

Augusta Arsenal Napoleon number 95, located at Gettysburg. *(Ross Ramer)*

the Government, therefore, developed its own foundries, the primary ones being at Augusta, Macon, and Columbus, Georgia. The foundries were under the jurisdiction of the arsenals in those cities. Under Government control domestic-made cannon witnessed more uniformity and a gradual improvement in quality.

The Augusta Arsenal, one of the original facilities commandeered from the United States, was able to commence production before either Macon or Columbus. The Arsenal, under the command of Colonel George Washington Rains, was organized in great part in the city where suitable buildings were obtained. Casting of cannon was done at the Augusta Foundry & Machine Works, located at the corner of Jackson and Campbell, Adams and D'Antignac. Construction on the facility was started in September, 1861, machinery being purchased at Holly Springs, Mississippi, Columbus, Georgia, and Selma, Alabama.[7]

Production of bronze 12-pounder Napoleons began in late 1862, the first castings being pronounced by Gorgas as "excellent." [8] "These guns [Napoleons] are now being manufactured rapidly at the government foundry and machine works in the city of Augusta, Georgia," reported a newspaper in early 1863, "and some have already been made for trial." [9] When completed, the guns were tested with a charge of ten pounds of powder and by loading them to the muzzle with bolts.[10] Testing was done near Sand Bar Ferry on the river just below Augusta and the initial use of the range frightened some into thinking that the city was under attack.[11]

Raw materials for cannon production came primarily from the Ducktown copper mines outside Chattanooga, captured 6-pounder bronze guns, state quarries, and scrap metal. The guns were made of a metal recently invented by the Austrians and recommended by Confederate diplomat James Mason. Austrian metal was a composition of copper, tin, wrought-iron and zinc.[12]

In June, 1863, a correspondent of the Augusta *Constitutionalist* visited the plant and reported the following:

> MUNITIONS OF WAR
>
> At the polite invitation of Captain I. P. Girardy, Paymaster, we on yesterday visited the Arsenal of artillery just set on foot in our city. It is located on the Canal, and comprises the old Augusta Machine Works, and is now turning out a fine Napoleon twelve pounder daily. These guns are made of the new Austrian metal, highly approved in Europe, and which has been found superior to any heretofore used in this country. It consists, we believe, of copper, zinc, and iron, and is susceptible of a polish equal to the old brass pieces. Our own State quarries furnish all of the ore necessary, save the zinc, which is used in a limited quantity, and is readily supplied.
>
> We found Mr. Van Buren, the well-known pomologist, of Clarksville, Ga., in charge of the works as Superintendent, and his success in reducing experiments with an untried composition to a practicle basis, attests his skill and capacity for the responsible post he has assumed. Great exigentics always find men equal to their drafts on human capacity, and Mr. Van Buren is a conspicuous illustration of this principle. Col. Rains seems to have been successful in securing, from the apparently limited number of artisons in the South, men in every department equal to the duties required, and by his own skill, experience, and untiring devotion to the work, has set on foot [?], some of which will rank favorably with any on the continent.[13]

In 1863 the Arsenal was visited by European correspondent Fitzgerald Ross. "We also went over the Government cannon-foundry, which is under the superintendence of Colonel Rains," recalled Ross. "The Colonel informed us that he could turn out a Napoleon a-day here, but at present it was not necessary. . . . They were then making Napoleons of Austrian metal . . . and had already turned out seventy." [14] Some of the Napoleons were sent to the nearby Georgia Railroad Machine Shop for completion.[15]

By mid-1863 production at all Government foundries was proceeding as scheduled, yet the heavy demands from all quarters caused the modernization to drag. In August, 1863, the Army of Tennessee possessed one-hundred twenty-seven cannon, including only twenty-one Napoleons, seven 10-pounder Parrotts, and two 20-pounder Parrotts.[16] That percentage slowly increased and by the Battle of Chickamauga in September thirty-one Confederate batteries reported a total of one-hundred twenty-one guns, thirty-nine of which were Napoleons.[17] Yet, the 6-pounder gun and 12-pounder howitzer still dominated. Chickamauga netted the Confederates a badly needed victory after the setbacks at Gettysburg and Vicksburg. The fruits of victory included thirty-nine captured pieces of Federal artillery among which were four Napoleons and seven 10-pounder Parrotts.[18] Many of the remaining captured guns eventually found their way to Government foundries for recasting into light 12-pounders.[19]

By the fall of 1863 the Government was also faced with the responsibility of re-equipping the Army of Vicksburg exchangees. "I have been wholly unable to procure artillery as yet," wrote Brigadier General F. M. Cockrill from his Demopolis, Alabama, headquarters. "I can get none here and have been unable to ascertain where I can get guns, caissons, etc. to outfit two or three artillery companies. . . . I desire the 12-pounder Napoleon gun."[20] Stevenson's Division, also exchanged from Vicksburg, was reassigned to the Army of Tennessee. All three of the Division's batteries were equipped with 12-pounder Napoleons by the late fall of 1863.[21]

Chickamauga was soon cancelled out by the Confederate defeat suffered at Chattanooga. The Federals bagged some thirty-eight pieces of artillery, including nineteen Confederate-made 12-pounder Napoleons. Six of these were cast by Leeds & Company of New Orleans and the remainder probably by government foundries.[22]

Augusta Arsenal number 95, located at Gettsyburg. Right trunnion face reads "GOVERT FOUNDRY & MACHINE WORKS AUGUSTA." *(Ross Ramer)*

After the Chattanooga defeat the Army of Tennessee went into winter quarters at Dalton, Georgia. In December, 1863, several of the more depleted batteries were transferred to garrison duty in Atlanta.[23] Even so several batteries in the army had only two or three guns and two batteries, Bledsoe's Missouri and Anderson's Georgia, were without guns.[24] By late March, 1864, a fresh consignment of ten Napoleons had arrived at Dalton. Four of the Napoleons were assigned to Bledsoe's Battery, while Anderson's Battery received four 10-pounder Parrott rifles.[25]

In March, 1864, an inspection of the Army's artillery was made by General William N. Pendleton, Lee's artillery chief. Regarding the armament he concluded: "The 6-pounders, of which there are fifteen, are, in the present state of fire arms, nearly useless, if not indeed worse, employing and exposing as they do a number of men and animals, while they can scarcely accomplish anything against the more powerful guns or even the long range muskets of the enemy." He recommended their immediate substitution by 12-pounder Napoleons.[26] Four light 12-pounders arrived shortly thereafter and were issued to the Louisiana Washington Artillery (5th Company). On May 4, 1864, four new Augusta Napoleons, numbers 73, 74, 75, and 76, arrived at the Atlanta Arsenal and were issued shortly thereafter.[27]

The 12-pounder howitzers, according to Pendleton, were "scarcely more valuable" than the 6-pounders. Although he did see the need for a few howitzer batteries for use in broken wooded country, the proportion in the Army (25 percent) was obviously too large. He was assured by the Chief of Ordnance that "several rifle bat-

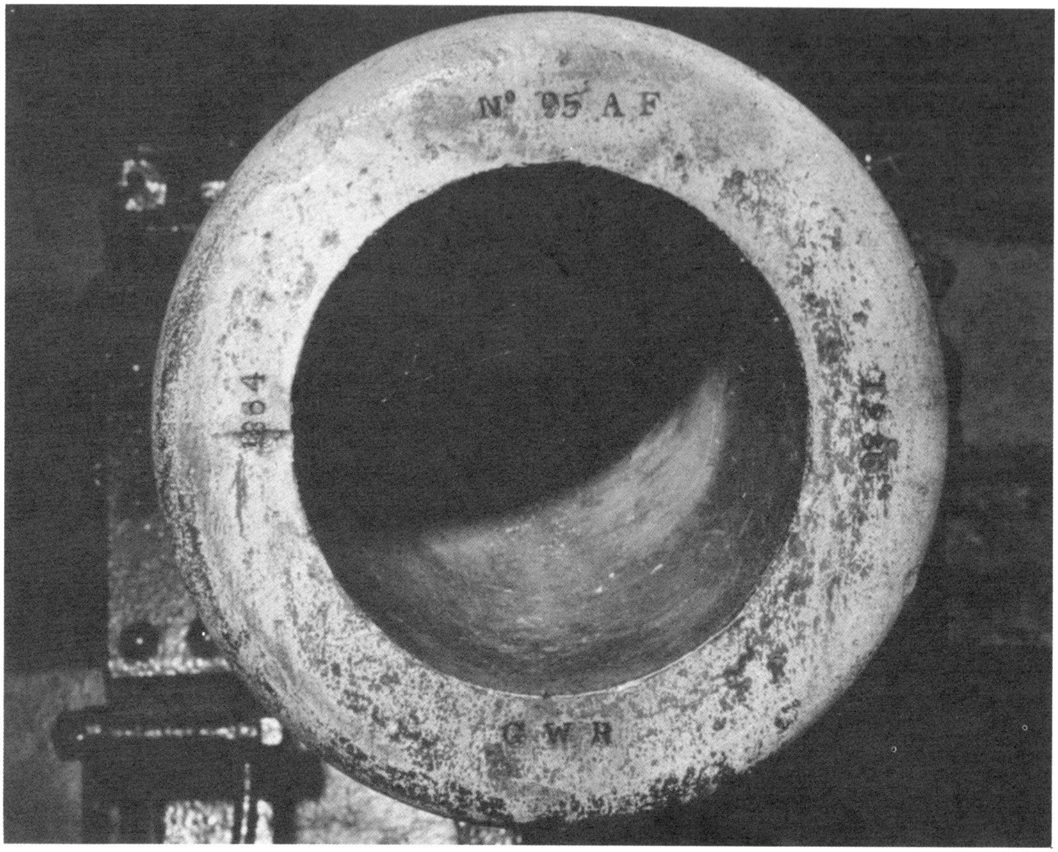

Top: Below:
Front and rear view of Augusta Arsenal Napoleon number 95, located today at Gettysburg. *(Ross Ramer)*

teries will be furnished in place of so many howitzers." Shortly thereafter two Alabama batteries—Emery's and Bellamy's—received four 10-pounder Parrotts and four English Blakely rifles respectively.[28]

The large number of Government Napoleons in service in the Army of Tennessee can be realized in a tabulation of cannon captured during the last months of the war. During the Nashville Campaign in latter 1864 the Confederates lost sixty-four field pieces, forty-five of which were of home manufacture. This included forty-two "light 12-pounders, rebel model." [29] In May, 1865, the Army of Tennessee surrendered 134 field pieces, namely, ninety-six C. S. guns, thirty-four U.S. guns, and four foreign pieces. The Confederate-made cannon included sixty-eight bronze light 12-pounders of "rebel manufacture." [30] Probably most, if not all, of these Confederate Napoleons were cast at Government foundries.

Dr. James Hazlett, who has done more work on the Napoleon gun than any other person, estimates total production at Augusta "in the neighborhood of 115–120." He bases this on the foundry numbers of forty surviving Augusta Napoleons. The represented low and high foundry number for each year was as follows: 1862—5–12; 1863—2–68; 1864—72–98. A new series of foundry numbers was started in 1863 but not in 1864.[31] Since this study, however, Dr. Hazlett has found gun number 115 dated 1864. This would probably push the total number cast to about 130. Colonel Rains stated that during 1863–64 some 110 field guns, mostly Napoleons, were issued at Augusta.[32] Almost all of these guns were issued to the Army of Tennessee.[33] It is possible that a few Parrott rifles were also cast at Augusta, although none survive.[34]

Thirty-three Augusta Napoleons survive at the Gettysburg park. The guns are marked with an "A.F." (Augusta Foundry) on the muzzleface and on the right trunnion face with: "GOVERT FOUNDRY & MACHINE WORKS AUGUSTA."

NOTES

[1] *O.R.*, XVII, pt. 2, 843.

[2] *Ibid.*, XXIV, pt. 2, 336, 418.

[3] *Ibid.*, XX, pt. 1, 768, 771, 856.

[4] *Ibid.*, XXIII, pt. 2, 762–763.

[5] *Ibid.*, 843.

[6] *Ibid.*, XXIII, pt. 2, 763.

[7] Florence F. Corley, *Confederate City: Augusta, Georgia, 1860–1865* (Columbia, S.C.: University of South Carolina Press, 1960), 50–52; Augusta *Constitutionalist*, September 11, 1861.

[8] Gorgas, "Notes on the Ordnance Department," 82.

[9] Memphis *Appeal*, March 27, 1863.

[10] Arthur James L. Fremantle, *Three Months in the Southern States: April–June, 1863* (New York: John Bradburn, 1864), 176.

[11] Corley, *Confederate City*, 52.

[12] Fremantle, *Three Months in the Southern States*, 176; Memphis *Appeal*, March 27, 1863; Gorgas, "Notes on the Ordnance Department," 82.

[13] Augusta *Constitutionalist*, June 7, 1863.

[14] Richard B. Harwell, ed., *Cities and Camps of the Confederate States by Fitzgerald Ross* (Urbana: University of Illinois Press, 1958), 142.

[15] Record of Receipts and Deliveries, Ordnance Office, Atlanta, Georgia, 1863, Chap. 4, Vol. 88, National Archives Record Group 109 (War Department's Collection of Confederate Records.)

[16] *O.R.*, XXX, pt. 4, 494.

[17] Compiled from the reports of battery commanders at the Battle of Chickamauga. See *O.R.*, XXX, pt. 2.

[18] *Ibid.*, XXX, pt. 1, 237. For a somewhat different statement of returns see *O.R.*, XXX, pt. 2, 41–42.

[19] *Ibid.*, XXX, pt. 4, 739.

[20] *Ibid.*, 669–670.

[21] *Ibid.*, XXXI, pt. 3, 826.

[22] *Ibid.*, XXXI, pt. 2, 99.

[23] Ibid., XXXI, pt. 3, 790, 821.

[24] Ibid., XXXI, pt. 2, 826–827.

[25] Ibid., XXXII, pt. 3, 687.

[26] Ibid., 685.

[27] Ibid., 731; Record of Receipts & Issues at the Atlanta Armory, December, 1863-June, 1864, Chap. IV. Vol. 17, pp. 222, National Archives Record Group 109, War Department's Collection of Confederate Records. (Information courtesy of Sydney Kerksis and Edwin Olmstead).

[28] *O.R.*, 731.

[29] Ibid., XLV, pt. 1, 48.

[30] Ibid., XLVII, pt. 1, 185–186.

[31] Hazlett, "The Confederate Napoleon Gun," 108.

[32] Corley, *Confederate City*, 114.

[33] G. W. Rains, *History of the Confederate Powder Works* Augusta: Chronicle and Constitutionalist Press, 1882), 30.

[34] Augusta *Constitutionalist*, January 27, 1864.

CHARLESTON ARSENAL

The Charleston Arsenal was siezed by South Carolina authorities on December 30, 1860. It was soon transferred to the control of the Confederate Government and greatly expanded. Between March and November, 1863, the Arsenal issued 16,000 heavy artillery projectiles, repaired nearly 10,000 small arms, fabricated three million small arm cartridges, and, toward the end of the year, turned out 2000 metallic friction primers a day. Heavy guns were also banded and repaired and heavy gun carriages constructed.[1] At that time 329 men between the ages of 16 and 60 were employed and organized into a defense battalion of five companies.[2]

Some cannon were also cast at the Arsenal, although few definite facts are known. On May 21, 1863, Colonel A. J. Gonzales addressed General Beauregard that "as soon as Napoleon guns are procured, of which four will soon be cast at the Charleston Arsenal, I will have the honor to earnestly advocate . . . the formation of batteries of horse artillery, with four Napoleons each. . . ."[3]

A possible further reference to Arsenal-made Napoleons was an August, 1863, statement that "the 12-pounder Napoleon gun, on board the *Juno,* will be returned to the arsenal for the purpose of having the sight put on it, and, when ready for service, will be issued, together with the other 12-pounder Napoleon, to Captain DePass."[4]

There are two surviving bronze Charleston Arsenal 12-pounder Napoleons. One is dated 1863 and the other, foundry number 14, is dated 1864. Both are inscribed on the right trunnion face: "CHARLESTON ARSENAL." Dr. James Hazlett conjectures that the guns were probably manufactured in a local foundry, sub-contracted by the Arsenal.[5] The probability is, however, that the guns were cast at the Arsenal. There were only two private foundries in Charleston which had the capacity to produce cannon—J. M. Eason & Brother and Cameron & Company. The records of these firms give no indication that any work of this type was done.[6]

Dr. Hazlett estimates maximum production at Charleston at 20–30 pieces, probably all 12-pounder Napoleons. Since the 1864 gun is number 14, however, it seems probable that production was below 20.

NOTES

[1] Columbus *Daily Sun,* December 6, 1863.

[2] Charleston *Mercury,* July 15, 1863.

[3] *O.R.,* XIV, 946.

[4] *Ibid.,* XXVIII, pt. 2, 284.

[5] Hazlett, "The Confederate Napoleon Gun," 108.

[6] J. M. Eason & Brother and Cameron & Company, "Citizens File."

The Charleston Arsenal *(Library of Congress)*

COLUMBUS ARSENAL

With the fall of New Orleans in April, 1862, the Baton Rouge Arsenal was considered in an exposed position. The facility was thus moved to Columbus, Georgia, where several private establishments were leased for the manufacture of gun carriages and leather goods.[1] The Columbus Iron Works, a private foundry, was already well established and engaged in cannon production for the Army. Upon the lease of that facility to the Navy in June, 1862, Army ordnance responsibility was shifted to the Columbus Arsenal, Captain F. C. Humphreys commanding. It was March, 1863, before construction on the Government Foundry was completed.[2] By early May it was reported that production was underway and that the foundry had the capacity to produce two guns per week.[3]

Columbus Arsenal Napoleons were channeled to both the Army of Tennessee and the Army of Mississippi.[4] Four Columbus Napoleons, numbers 25, 26, 27 and 28 were received at the Atlanta Arsenal on March 1, 1864. These were, most likely, issued to the Army of Tennessee. Three of the guns, numbers 26, 27, and 28 survive at Gettysburg.[5] Some were also shipped to the Mobile defenses. In August, 1863, Major Henry Myers, Chief of Ordnance of the Department of the Gulf, reported that two four gun Napoleon batteries had been ordered and would soon arrive at Mobile.[6] These guns were probably cast at Columbus.

Dr. Hazlett has found twenty-seven surviving Columbus Napoleons, numbers 3-23 dated 1863 and 26-52 dated 1864. Napoleon number 52 has a right rimbase number of 71. This would indicate that at least seventy-one guns were cast at Columbus, fifty-two of which were Napoleons. The difference of nineteen indicates that several calibers of guns were cast and that perhaps some

The only known surviving Columbus Arsenal 9-pounder. *(Tarry Beasley)*

Markings on a Columbus Arsenal Napoleon. *(Tarry Beasley)*

castings were porous and were rejected. It is known that at least one caliber of gun was cast other than Napoleons. Several years ago one of the authors came into possession of an 1863 bronze Columbus Arsenal 9-pounder, the only one known to survive. Markings indicate that it was the thirty-third gun cast at Columbus, the fifth of its kind. Dr. Hazlett estimates the total number of Napoleons cast at Columbus at between sixty and seventy. Accepting the low number (which is more nearly accurate) and adding about twenty (the approximate difference in rimbase numbers) would put the total number of guns cast in the neighborhood of eighty.

Most surviving Columbus Napoleons are to be found at the Gettysburg National Military Park. The weapons are inscribed on the muzzleface, for example: "No. 35 F.C.H. C.S. ARSENAL COLUMBUS, GEO. 1864 1221".

Columbus Arsenal Napoleon number 45. The rimbase number indicates that this was the 64th gun cast at Columbus. *(Ross Ramer)*

NOTES

[1] Columbus *Daily Sun,* June 4, 1862.

[2] Charleston *Courier,* March 11, 1862.

[3] Richmond *Examiner,* May 13, 1863.

[4] O.R., XXXVIII, pt. 4, 656–657.

[5] Record of Receipts & Issues at Atlanta Armory, December 1863–June, 1864, Chap. IV, Vol. 18, National Archives Records Group 109, War Department's Collection of Confederate Records. (Information courtesy of Sydney Kerksis and Edwin Olmstead).

[6] O.R., XXVI, pt. 2, 161.

[7] Hazlett, "The Confederate Napoleon Gun," 108.

The Macon Arsenal Cannon Foundry *(National Archives)*

MACON ARSENAL

In April, 1862, the Confederate Government purchased the Findlay Iron Works in Macon, Georgia. Thereafter the foundry was operated under the auspices of the Macon Arsenal, Captain Richard M. Cuyler commanding. Until bought out by the Government the establishment, owned and operated by J. D. and C. N. Findlay, turned out railroad car castings, car wheels, and steam engines.[1] Only eleven hands were originally employed—1 negro molder, 1 negro engineer, 1 striker, 6 machinists, 1 negro machine blacksmith, and 1 shipping clerk. It was estimated, however, that an additional 40–50 hands could be procured in the Macon vicinity.[2]

Plans were made during the summer of 1862 for the casting of bronze 6-pounders and 12-pounder howitzers as described by the Ordnance Manual, although certain officers urged the casting of 12-pounder Napoleons even at that early date. Captain Cuyler also requested that the Ordnance Department send the drawings and specifications of the 3-inch rifle, iron 6-pounder, and iron 12-pounder howitzer.[3] Possibly a few samples were turned out before the November, 1862, circular from Gorgas restricting future production to Napoleons and Parrotts.[4]

Initial raw material shortages plagued production capabilities. Upon his arrival Cuyler petitioned Richmond authorities for permission to appeal to local churches and plantations for bells to be melted down. On May 7, 1862, Gorgas responded: "Captain, Your letter of the 30th. has been received and contents noted. Get your works at Macon into operation as soon as possible, and get up 12 4-gun batteries complete. Bells may be called for to be replaced at the close of the war." [5] Response was soon forthcoming.

Production of bronze 12-pounder Napoleons began at Macon in early 1863 and by mid-June two complete batteries had been turned out. "Another beautiful battery of twelve pound bronze Napoleon guns from the Macon Arsenal was on the streets yesterday. . . ," reported the Macon *Telegraph*. "Both these batteries will be the pride of our artillerymen. The guns are beautifully shaped, and the carriages and caissons thoroughly made and handsomely painted. We understand the old 6-pounder field pieces have been pretty much abandoned in our service." [6]

Macon Napoleons were channeled to the Army of Tennessee and the Army of Mississippi. An 1863 (month unknown) listing of the guns in Walker's Division, Army of Mississippi, revealed that a battery of Macon Napoleons, foundry numbers 5, 6, 9, and 10, was in use.[7] All four of these guns survive today at the Gettysburg park. In the defeat suffered by the Confederates at Chattanooga, some thirty-eight field pieces were lost, nineteen of which were Confederate-made Napoleons.[8] The photograph on page 72 shows some of the guns captured at Missionary Ridge. A close study will reveal that the gun carriages are stenciled: "MACON ARSENAL 1863." Few additional records relating to specific Macon guns are available. It is known, however, that on April 6, 1864, the Atlanta Arsenal received four Macon 12-pounder Napoleons, foundry numbers 41, 42, 43, and 44.[9] Number 41 now survives at Rock Island and number 43 survives at Gettysburg. On June 3rd, a Macon 10-pounder Parrott number 11 was issued through the Arsenal.

While the Augusta Foundry was largely responsible for

CAPTURED MACON NAPOLEONS

These Confederate 12-pounder Napoleons were captured at the Battle of Missionary Ridge. Stenciled on the gun carriage trail is the inscription: "Macon Arsenal 1863". *(Library of Congress)*

modernizing the field artillery of the Army of Tennessee, it was the Macon and Columbus Arsenals which largely assisted in the artillery transition of the Army of Mississippi. In the spring of 1864 three batteries in Loring's Division, Point Coupe Louisiana, Barry's Tennessee, and Charpentier's Mississippi, turned in their six 6-pounder guns and six 12-pounder howitzers for twelve Napoleons. In French's Division, Guibor's Missouri Battery, exchanged from Vicksburg, was re-equipped with four Napoleons.[10] By May 1, 1864, the authorities planned to exchange the remaining 6-pounders in that Army with Napoleons and Parrots, including four 10-pounder Parrotts from Macon.[11]

Between March, 1863, and June, 1864, the Macon Arsenal made definite deliveries of seventy-two cannon as follows: [12]

An 1864 Macon Arsenal 20-pounder Parrott at Gettysburg. *(Tarry Beasley)*

MONTH OF DELIVERY	DELIVERED	ARSENAL
March 1863	2 12-pounder Napoleons	Macon
April	3 12-pounder Napoleons	"
May	8 12-pounder Napoleons	"
June	4 12-pounder Napoleons	"
July	3 12-pounder Napoleons	"
August	5 12-pounder Napoleons	"
	2 10-pounder Parrotts	
September	2 10-pounder Parrotts	"
October	4 12-pounder Napoleons	"
	1 20-pounder Parrott	
November	7 12-pounder Napoleons	"
	1 30-pounder Parrott	
December	4 10-pounder Parrotts	"
February 1864	3 12-pounder Napoleons	"
March	4 12-pounder Napoleons	"
April	4 12-pounder Napoleons	"
	3 10-pounder Parrotts	
	3 20-pounder Parrotts	
May	1 10-pounder Parrott	"
	1 20-pounder Parrott	
	4 12-pounder Napoleons	
June	2 12-pounder Napoleons	"
	TOTAL CANNON DELIVERED 72	

During that same time period the Arsenal also issued 16 siege carriages, 55 field carriages, 41 caissons, 12 forges, and 15 battery wagons. Inquiries were made in the summer of 1864 to ascertain if any heavy columbiads could be cast at Macon.[13] There is no indication that any were produced.

Dr. Hazlett has found thirty surviving Macon Napoleons, numbers 1–37 dated 1863 and numbers 40–49 dated 1864. He states: "Assuming that about the same numbers were made here as in other foundries in 1864, this would give a total of about seventy pieces. This assumption may seem too high, but one can never be sure that the last serial number of 1864 has been recovered, or even close to the last one. . . ." [14] Dr. Hazlett's estimate of seventy would indeed seem high. The highest surviving Macon Napoleon is number 49 with a right rimbase number of 65. This would indicate that at least sixty-five guns had been cast up to that time, forty-nine of which were Napoleons. It is known that on June 30, 1864, Macon Napoleon number 56 was sent through the Atlanta Arsenal en route to the Augusta Arsenal. That is the highest known Napoleon number to have been cast. Add to that eighteen the known number of Parrotts to have been cast, and the total would be seventy-four. Surviving weapons, most of which can be found at the Gettysburg park, are inscribed on the muzzleface, for example: "MACON ARSENAL 1863 NO. 19 T.A.S. 1180." The "T.A.S." stands for T. A. Sengstack, one of several inspectors at the foundry.

A Macon Arsenal Napoleon with markings at the Gettysburg National Military Park. *(Tarry Beasley)*

Macon Arsenal Napoleon number 11, survives at Gettysburg. *(Tarry Beasley)*

Three Macon Parrotts, one 10-pounder and two 20-pounders, are also at Gettysburg. The two 20-pounders were cast in 1864 and weigh 1654 and 1660 pounds respectively. The guns are inscribed on the right trunnion face: "MACON ARSENAL."

NOTES

[1] "What We Are Gaining By The War," *Debow's Review,* VIII (April 1862), 79.

[2] Macon Arsenal Letters Received, Chapter 4, Vol. 4, National Archives Record Group 109 (War Department's Collection of Confederate Records). Hereafter cited as "Macon Arsenal Letters Received."

[3] Briggs to Cuyler, May 31, 1862, Ibid.

[4] *O.R.,* XXI, 1047.

[5] Gorgas to Cuyler, May 7, 1862, "Macon Arsenal Letters Received."

[6] Macon *Telegraph,* June 13, 1863.

[7] Listing of Cannon in Walker's Division, Army of Mississippi, 1863 (month unknown), Mississippi State Archives.

[8] *O.R.,* XXXI, pt. 2, 99.

[9] Record of Receipts and Issues at the Atlanta Armory, December 1863–June 1864, Vols. 17–18, National Archives Record Group 109 (War Department's Collection of Confederate Records). Information courtesy of Edwin Olmstead.

[10] *O.R.,* XXXII, pt. 2, 519–520.

[11] *Ibid.,* XXXVIII, pt. 4, 656–657.

[12] Daily Report of Ordnance and Ordnance Stores on Hand, Macon Arsenal, 1863–64, Chapter 4, Vol. 58, National Archives Record Group 109 (War Department's Collection of Confederate Records).

[13] *O.R.,* XXXV, pt. 2, 605.

[14] Hazlett, "The Confederate Napoleon Gun," 108.

Ruins of the Selma Naval Gun Foundry. *(Mrs. Bert Neville)*

SELMA NAVAL GUN FOUNDRY

In February, 1862, Colin J. McRae, a prominent Mobile cotton merchant and an Alabama representative to the Provisional Congress, approached the War Department concerning the prospect of building a cannon foundry and rolling mill in Selma, Alabama. The Government rejected the proposition but McRae soon obtained the necessary funds from private sources. By March, 1862, contracts for cannon had been arranged with the Government, the first gun to be cast by September 1st.[1]

A number of construction and labor difficulties were encountered from the beginning and the facility was never completed while McRae was in charge. In early 1863 the Richmond Government requested that he undertake an important financial mission to Europe in behalf of the Confederacy. McRae agreed on the condition that he could be relieved of the Selma Foundry without loss to himself. Confederate authorities favored Cartersville, Georgia, over Selma as the site for a national foundry but, as stated Maurice Melton, "McRae and his foundry were a package deal." Negotiations were thus begun and by March 1st both the War and Navy Departments had agreed to operate the works on a joint basis.[2]

On February 20, 1863, Colonel George W. Rains was ordered to assume command at the works on behalf of the Army and Navy. The choice was an unfortunate one. Rains had long deplored Selma as the location for a large ordnance establishment and as commander his views did not change. He soon began advocating that the Navy assume total responsibility. His criticisms were effective, for on June 1, 1863, the Navy Department assumed complete control, although still committed to supplying the needs of both services.[3]

Commander Catesby ap R. Jones of Virginia was the Naval officer placed in charge. Jones, formerly of the United States Navy, had been the executive officer aboard the ironclad *Virginia* during its famed encounter with the *Monitor*. Jones was an intelligent and competent officer and proved a capable commander.[4]

The condition of the foundry upon his arrival was far from encouraging. Although the facility had been under construction for well over a year, not one gun had been cast. Furthermore, much of the work started by McRae had to be redone. The gun pit, for example, was located in an area in which there was a constant seepage of water.[5]

Jones was directed to concentrate his attention upon the casting of Brooke guns. The Brooke was designed by John M. Brooke, a naval scientist and head of the Navy Bureau's Ordnance and Hydrography Department. Brookes had been cast at the Tredegar Iron Works since the fall of 1861 and were held in high regard by Confederate authorities. During 1863 a number of Tredegar Brookes either burst or were in need of repair work after delivery. Yet, all things considered, Brookes were extremely effective cannon, by far the most powerful in the Confederate arsenal. The Brooke certainly had a more credible performance than large caliber Northern-made Parrotts, which were notorious for bursting at the muzzle. Brookes had another advantage—they had been proven effective against ironclads of the Monitor class.[6]

Casting commenced in the summer of 1863. The choice of a proper iron with which to cast heavy guns was an important one. The Navy recommended iron from nearby Bibb County. Jones first set about to cast a number of experimental 6-pounders to test the durability of the iron and to discover the proper method for casting. The first heavy gun, a 7-inch Brooke, was cast July 30th. The experiment proved a total failure. The furnaces failed and the fire brick burst. New brick had to be gathered and the furnace rebuilt. While construction was taking place Jones sent his foundry superintendent, John Peacock, to the Tredegar Iron Works to study casting techniques.[7]

By the end of August two additional 7-inch Brookes had been cast. The weapons could not be completed, however, since the special lathe to turn the trunnions had not arrived from Richmond. Nor would it. Jones finally had to have one of his mechanics construct one.[8]

Completing a 7-inch double-banded Brooke rifle, weighing in excess of 14,000 pounds, was a lengthy task. The process began by tapping metal (usually No. 1 or No. 2 cold blast pig from Bibb County) from several large reverberating furnaces. When the metal was totally fused (requiring about two hours) it was poured into the gun mold and the mold placed into a pit to cool for about eight days. The casting was then sent to the boring mill. The various stages of boring sometimes required 900 hours. With boring completed the gun was next rifled (about 13 hours), turned in a lathe (about 31 hours), and the trunnions finished (about 20 hours).

In the next step the inner and outer bands of the Brooke were attached, a process requiring about 51½ hours. The band was expanded by heating until it could be slipped over the breech of the cannon. When it cooled it shrunk tightly into place. The outside of the first band was then turned and the process repeated with the second band. The weapon was then sighted, the cascabel block finished, and the vent bored. Working night and day a Selma 7-inch Brooke could be completed in 2½–3 months.[9]

During the experimental casting of cannon, the production of shot, shell, and machine parts went on.

Colin J. McRae (*Ala. State Archvies)*

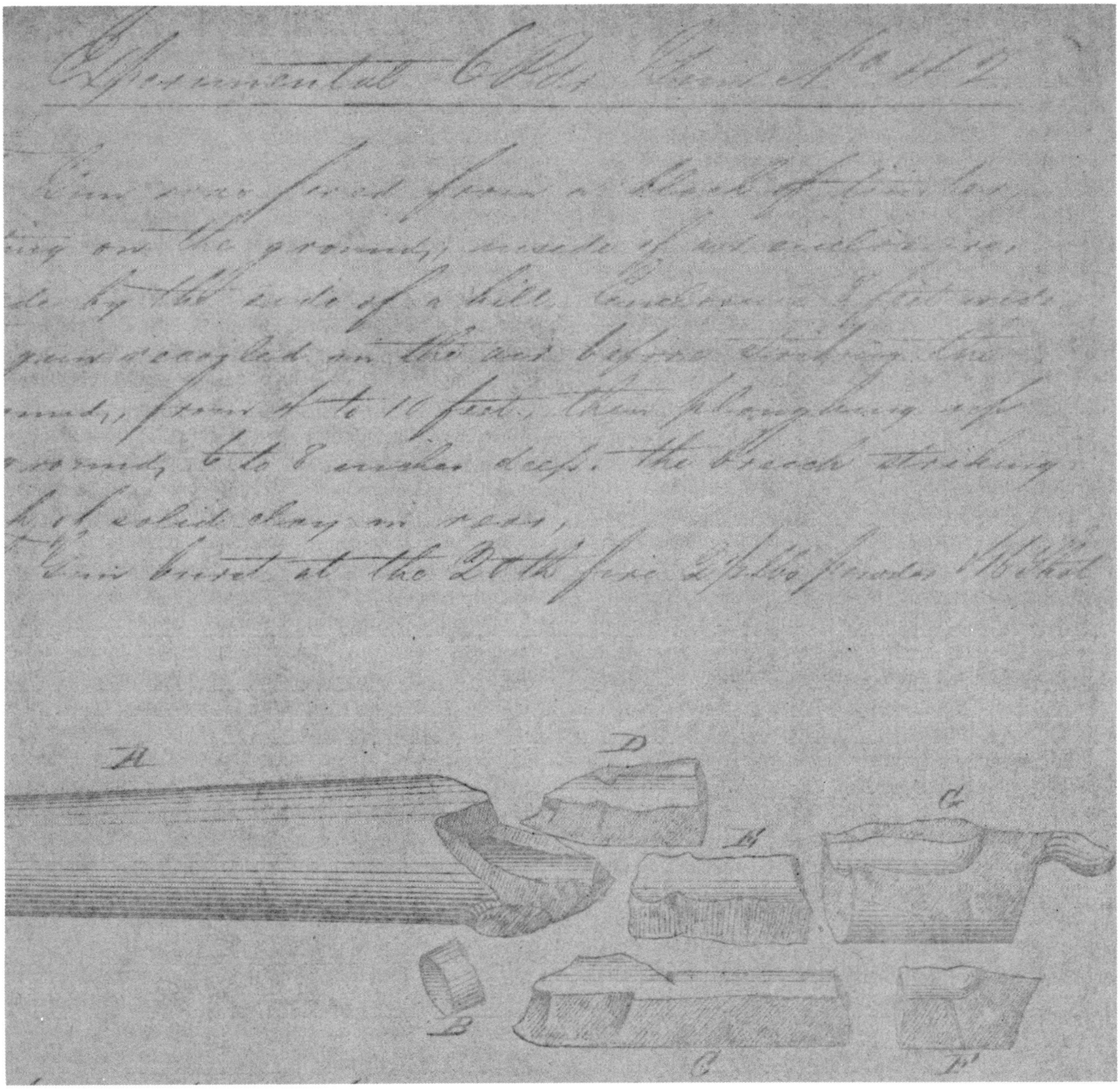

One of twenty experimental 6-pounders cast at the Selma Foundry. All of the guns were purposely burst during firing tests to ascertain the strength of the metal. *(National Archives)*

Throughout the spring of 1863 the plant supplied most of the large caliber shot used at Vicksburg and Port Hudson and many of those used at Charleston. Daily production of shells rose from 16 to 20, and later to 30. By the time casting of heavy guns commenced the plant had already furnished 4300 general castings, 3200 large caliber projectiles, 4300 pounds of spherical case bullets for the Army, plus 400 shot and shell for the Navy.[10]

By the end of September, 1863, the first of the 7-inch Brookes was half finished. It was bored and vented but still unrifled, unbanded, and the trunnions incomplete; there was still no trunnion lathe. On November 1st it was tested in an unfinished state to ascertain the quality of the metal. The test proved successful and the metal excellent.[11]

The Naval Gun Foundry, as the facility was now called, continued to experience frustrating failures throughout the fall of 1863. Three Brookes were cast between October 8–19, all of which were condemned. These difficulties were attributable to inexperience in casting. Several guns were ruined by cavities caused by overheating the metal.[12]

Some progress had been achieved at the foundry by November, although the facility was far from complete. In that month Jones reported that the gun pit had been bricked, a reservoir formed, two air furnaces braced, and sheds built to cover them. The end walls and roof of the foundry were completed. At the machine shop and boring mill the roof was braced and strengthened, tools and bits completed, gun flasks made, and work begun on a specially constructed railroad car for hoisting and carrying heavy guns. Some furnaces were completed at the blacksmith shop for heating the bands of the Brookes.[13] The next month the trunnion lathe was completed and put in operation and a machine for bending the Brooke bands arrived.[14]

In November, Commander Brooke pressed Jones to complete several Brookes for a "contemplated operation" by Admiral Buchanan, commanding the C. S. naval fleet at Mobile.[15] The projected operation was the completion of the powerful ironclad *Tennessee.* The Tredegar works supplied four 6.4-inch Brooke rifles for the vessels, but the bow and stern guns were to be two Selma 7-inch guns. On January 8, 1864, a 7-inch gun (S-5) was loaded aboard the boat *Southern Republic* for Mobile Bay. "The gun is our first attempt," Jones wrote Buchanan. "I therefore invite criticism. Those guns that follow may be improved by it." [16] Two weeks later two additional 7-inch guns (S-10 and S-19) were delivered to Mobile and a fourth (S-8) was shipped on January 30th. The iron bands, elevating screws, and gun carriages were supplied by the naval works at Charlotte and Atlanta.[17]

By January, 1864, the Selma Naval Gun Foundry was nearly completed. In late December an explosion in one of the gun pits resulted in the burning of several buildings but damage was slight to the foundry.[18] The completed works included offices, gun foundry, machine shop, boring mill, molding shop, melting furnaces, pattern shop, puddling furnace, and blacksmith shop. Some of the machinery had been previously purchased by McRae in New Orleans, Atlanta, and Montgomery, some shipped from the abandoned Pensacola Navy Yard, some shipped through the blockade, and some made at the foundry itself.[19]

Casting difficulties continued to be experienced throughout the early months of 1864. In January alone three Brookes were condemned after casting.[20] "The metal we use is very liable to cavities," Jones wrote Buchanan in the spring. ". . . The metal with which we are furnished for guns differs so much that the test gun is no guide to us." [21]

In early March Jones received distressing news from Mobile. On February 29th a Selma 7-inch Brooke (S-19) burst during the heat of a bombardment at Fort Powell. The weapon, which had been fired at various elevations, became overheated and burst at the muzzle on the fifty-fifth round. The bursting was attributed to the premature explosion of a shell, but the commanding officer later reported that "upon careful examination of the fracture I believe that the cause was a defect in the gun itself." [22]

The Navy Department continued to press Jones for heavy guns with which to arm its vessels under construction. In January two 7-inch Brookes had been delivered to Mobile for the ironclad *Tennessee.* In early May Jones wrote that four 7-inch guns had been cast for the *Nashville.*[23] In July two 7-inch guns were delivered to South Carolina to be placed aboard the gunboat *Peedee.* The gunboats *Gaines, Morgan,* and *Selma* in Mobile Bay were the recipients of about four Selma 6.4-inch and 7-inch guns. In September the first of two 7-inch guns were delivered to Columbus, Georgia, to be placed aboard the *Chattahoochie.* In February and March, 1865, five Selma Brookes (two 6.4-inch and three 7-inch) were shipped to Columbus for the *Muscogee.*[24] About one-third of all 6.4-inch and 7-inch guns completed at Selma were placed aboard naval vessels.

Jones was also committed to meet the needs of the Army. Prior to the spring of 1864, production had centered on the 6.4-inch and 7-inch Brooke rifle and the 8-inch Brooke smoothbore, which was the 6.4-inch rifle reamed out to 8 inches with no change in the exterior. The Tredegar Iron Works had cast a few 10-inch and 11-inch Brooke smoothbores in the fall and winter of 1863–64 and Jones was most anxious to produce some of the large caliber weapons for harbor defense. He was convinced that these guns could stop the Union ironclads.[25]

Selma mechanics had been preparing a 10-inch gun pattern since late 1863, but it was May 3rd before the first 10-inch was poured. Over a week later Jones in-

Catesby ap. R. Jones *(Ala. State Archives)*

A row of Selma and Tredegar Brookes at the Washington Navy Yard. *(Tarry Beasley)*

formed Mobile army officers that the gun was still in the pit because its tremendous weight, 24,000 pounds, had played havoc with the hoisting equipment.[26] It was July 29th, nearly three months later, before the weapon was bored, doublebanded, and shipped to Mobile. The first 11-inch smoothbore was cast on July 2nd and a second on August 10th. One was eventually shipped to Wilmington in September and the other to Charleston in November. The Charleston gun was delivered only after lengthy delays due to poor rail transportation.[27]

Production of heavy guns at Selma was restrained due to two factors—the shortage of skilled labor and proper gun iron. "I regret that we cannot turn out guns faster," Jones wrote Buchanan on March 1st. "You are aware of the cause, want of mechanics. . . . Anything that you can do to increase our force of molders, machinists and blacksmiths will enable us to furnish you more rapidly." [28] In May Jones elaborated the problem. "We now cannot turn out more than one gun per week, but with a proper number of mechanics could turn out three a week, and in a few months one a day." [29] A few weeks later Jones remarked that improvements had been made in the machinery for boring and turning the guns and bands, "but without more blacksmiths we cannot finish the guns faster than heretofore. I find it useless to represent such things at Richmond. It produces no effect." [30]

Slave labor was used extensively at the plant as the war progressed. Negroes were usually hired from local planters for a year and employed primarily in cutting wood for charcoal and other unskilled tasks.[31] By February 1, 1865, no less than 300 of the 450 hands employed were slaves.[32] The work load was grinding and the hours long for both white and black. Mechanics labored from 8 A.M. to 9 P.M., six days a week, and occasionally on Sundays.[33]

Even had a sufficient number of mechanics been available, production would still have been limited because of the shortage of materials—wood, coal, and iron.[34] In March, 1864, Jones informed Major Hunt of the Niter and Mining Bureau: "I am constantly receiving urgent demands from all quarters. . . . Our capacity to furnish the guns is, of course, much circumscribed by the limited supply of iron, none having been received here this month." [35]

By the end of 1864, the deteriorating rail system further complicated the shipment of gun metal. By that time, however, a large enough supply had been stockpiled so that new orders could be filled from the existing stock.[36] Between September and December, 1864, the foundry cast seventeen Brooke guns, only one of which was condemned.[37]

Siege operations around Mobile also became a primary concern for the works in the summer and fall of 1864. Between July and September, 1864, nineteen 24-pounder coehorn mortars were cast, the first three of which were condemned and the others shipped to the Gulf City.[38] Shortly before Christmas orders were taken for iron 12-pounder Napoleons and 20 and 30-pounder Parrott rifles. Production of Napoleons and 20-pounder Parrotts never developed, but on December 8th the first 30-pounder was cast. The gun was condemned. By January 10th, however, two 30-pounders were being machined, and ten more were in or ready for the boring mill. By the end of February, 1865, five 30-pounders had been shipped to the Mobile defenses.[39]

During the course of its existence the Selma Naval Gun Foundry turned out no less than 102 Brooke guns of various calibers, nineteen coehorn mortars, twenty experimental 6-pounders, and at least twelve 30-pounder Parrott Rifles. A number of guns, some condemned and others completed and awaiting shipment, were at the works when Selma was captured on April 2, 1865. Much of the machinery at the plant was destroyed by the workmen the previous night. Portable tools and machines and some cannon were rolled into the river. The gun foundry and buildings were later destroyed by Federal troops.[40]

Eleven Selma Brookes are known to survive as follows: [41]

TYPE	NUMBER	PRESENT LOCATION
7-in.	S–5	Washington Navy Yard
7-in.	S–10	Washington Navy Yard
8-in.	S–26	Gainesville, Alabama
7-in.	S–76	Charleston, South Carolina
7-in.	S–81	Columbus, Georgia
10-in.	S–87	Columbus, Georgia
7-in.	S–89	Fort Morgan, Alabama
7-in.	S–95	Choctaw Bluff, Alabama
6.4-in.	S–96	Jackson, Alabama
11-in.	S–97	Columbus, Georgia
7-in.	S–111	Columbus, Georgia

The 6.4-inch and 7-inch cannon are rifles and the 8-inch, 10-inch, and 11-inch weapons are smoothbores. All survivors are double banded. Brooke guns vary slightly in their inscriptions. Generally, however, the caliber of the gun is inscribed on both trunnion faces, "C. ap R. J." (Catesby ap R. Jones) and the date on the right trunnion face, and the foundry number on the breech. Occasionally the weight is also inscribed on the left trunnion face.

NOTES

[1] William N. Still, "Selma and the Confederate States Navy," *The Alabama Review,* XV (Jan. 1962), 20–21.

[2] *Ibid.,* 21–22; Maurice Melton, "The Selma Naval Ordnance Works," *Civil War Times Illustrated,* XIV (Dec. 1975), 19; *O.N.R.,* Series 2, II, 231–232.

[3] Still, "Selma and the Confederate States Navy," 23.

[4] *Ibid.*

[5] *Ibid.,* 24.

[6] Hazlett, "The Parrott Rifle," 28; *O.N.R.,* Series 2, II, 639, 641 and Series 1, XXI, 874.

[7] Melton, "The Selma Naval Ordnance Works," 22.

[8] *Ibid.*

[9] Record of Operation, Selma Naval Gun Foundry, Vols. I and II, National Archives Record Group 45 (Naval Records). Hereafter cited as "Selma Foundry Record of Operations"; *O.N.R.,* Series 1, XX, 857.

[10] Melton, "The Selma Naval Ordnance Works," 22.

[11] *Ibid.,* 22.

[12] *Ibid.,* 23; "Selma Foundry Record of Operations."

[13] *O.N.R.,* Series 2, II, 548–549.

[14] *Ibid.,* Series 1, XX, 857 and XXI, 859.

[15] *Ibid.,* XX, 851.

[16] *Ibid.,* XXI, 863.

[17] Melton, "The Selma Naval Ordnance Works," 24; "Selma Foundry Record of Operations."

[18] *O.N.R.,* Series 1, XX, 858.

[19] Still, "Selma and the Confederate States Navy," 25; Melton, "The Selma Naval Ordnance Works," 19.

[20] "Selma Foundry Record of Operations."

[21] *O.N.R.,* Series 1, XXI, 892.

[22] *Ibid.,* 881, 883.

[23] *Ibid.,* 896.

[24] Walter W. Stephens, "The Brooke Guns From Selma," *The Alabama Historical Quarterly,* XX (Fall 1958), 465–466; "Selma Foundry Record of Operations."

[25] *O.N.R.,* Series 1, XXI, 910.

[26] *Ibid.,* 899.

[27] "Selma Foundry Record of Operations"; Melton, "The Selma Naval Ordnance Works," 30.

[28] *O.N.R.,* Series 1, XXI, 879.

[29] *O.R.,* Series 4, III, 523.

[30] *O.N.R.,* Series 1, XXI, 897.

[31] Still, "Selma and the Confederate States Navy," 29.

[32] Brooke to Mallory, February 1, 1865, Office of Hydrography Letterbook, National Archives Record Group 45 (Naval Records).

[33] Melton, "The Selma Naval Ordnance Works," 29.

[34] *Ibid.*

[35] *O.N.R.,* Series 1, XXI, 882.

[36] Melton, "The Selma Naval Ordnance Works," 30–31.

[37] "Selma Foundry Record of Operations."

[38] *Ibid.*

[39] Melton, "The Selma Naval Ordnance Works," 31.

[40] Still, "Selma and the Confederate States Navy," 36–37.

[41] "Brooke Digest," an unpublished paper by Edwin H. Olmstead.

WEAPONS CAST AT THE SELMA NAVAL FOUNDRY

NO.	TYPE	CAST	DELIVERED	DESTINATION
S–1	6-pdr.	July 21, 1863		Experimental
S–2	7-in.	30		Condemned
S–3	6-pdr.	Aug. 15		Experimental
S–4	7-in.	15		
S–5	7-in.	24	Jan. 8, 1864	Mobile
S–6	6-pdr.	29		Experimental
S–7	6-pdr.	31		Experimental
S–8	7-in.	Sept. 7	Jan. 30	Mobile
S–9	6-pdr.	11		Experimental
S–10	7-in.	12	Jan. 22	Mobile
S–11	7-in.	16		
S–12	6-pdr.	17		Experimental
S–13	6-pdr.	17		Experimental
S–14	6-pdr.	23		Experimental
S–15	6-pr.	23		Experimental

NO.	TYPE	CAST	DELIVERED	DESTINATION
S–16	7-in.	Oct. 8		Condemned
S–17	7-in.	10		Condemned
S–18	6.4-in.	19		Condemned
S–19	7-in.	24	Jan. 22	Mobile
S–20	8-in.	3	May 17	Mobile
S–21	7-in.	Nov. 28		
S–22	6.4-in.	30		
S–23	7-in.	Dec. 2		
S–24	7-in.	3	Feb. 17	Mobile
S–25	7-in.	8	Feb. 17	Mobile
S–26	8-in.	11	June 27	Mobile
S–27	7-in.	12	Feb. 27	Mobile
S–28	7-in.	15	July 26	Mobile
S–29	8-in.	21	April 1	Mobile
S–30	7-in.	24	April 17	Mobile
S–31	6.4-in.	24	March 4	Mobile
S–32	7-in.	Jan. 2, 1864	July 26	Mobile
S–33	8-in.	5	March 23	Mobile
S–34	6.4-in.	6	Sept. 28	Charlotte
S–35	8-in.	19	April 15	Mobile
S–36	7-in.	26	Sept. 8	Augusta
S–37	6.4-in.	26		Condemned
S–38	7-in.	27		Condemned
S–39	6.4-in.	Feb. 6		Condemned
S–40	6-pdr.	13		Experimental
S–41	6.4-in.	13	July 26	Mobile
S–42	6-pdr.	22		Experimental
S–43	6.4-in.	22	July 26	Mobile
S–44	7-in.	April 5	Aug. 19	Mobile
S–45	7-in.	11	July 3	Mobile
S–46	7-in.	12	July 3	Peedee, S.C.
S–47	6.4-in.	13	Aug. 19	Mobile
S–48	6.4-in.	16	July 15	Mobile
S–49	6-pdr.	20		Experimental
S–50	7-in.	20	July 13	Mobile
S–51	7-in.	23		
S–52	7-in.	27	July 29	
S–53	6.4-in.	29	July 13	Peedee, S.C.
S–54	10-in.	May 3	July 29	Mobile
S–55	6-pdr.	6		Experimental
S–56	7-in.	6	July 26	Mobile
S–57	7-in.	7	July 15	Mobile
S–58	7-in.	13	Aug. 15	Mobile
S–59	6.4-in.	14	Nov. 10	Wilmington
S–60	6-pdr.	16		Experimental
S–61	6.4-in.	16	Aug. 10	Mobile
S–62	10-in.	20	Feb. 24 (1865)	Mobile
S–63	7-in.	21	Sept. 7	Mobile
S–64	6-pdr.	24		Experimental
S–65	7-in.	24	Aug. 18	Mobile
S–66	7-in.	27	Aug. 7	Mobile
S–67	6.4-in.	28	Aug. 26	Mobile
S–68	7-in.	June 1	Aug. 2	Mobile
S–69	6.4-in.	2		Condemned

NO.	TYPE	CAST		DELIVERED	DESTINATION
S–70	7-in.		4	Aug. 24	Charleston
S–71	7-in.		7		Condemned
S–72	6.4-in.		8		Condemned
S–73	10-in.		10	Sept. 24	Wilmington
S–74	7-in.		14	Sept. 2	Charleston
S–75	6-pdr.		15		Experimental
S–76	7-in.		15	Sept. 19	Charleston
S–77	10-in.		25	Sept. 10	Mobile
S–78	6.4-in.		27	Sept. 15	Mobile
S–79	7-in.		28		Condemned
S–80	11-in.	July	2	Sept. 28	Wilmington
S–81	7-in.		6	Sept. 10	Columbus
S–82	6.4-in.		8		Condemned
S–83	7-in.		9		Condemned
S–84	6-pdr.		13		Experimental
S–85	6.4-in.		14	March 3, 1865	Columbus
S–86	7-in.		16	Nov. 10, 1864	Wilmington
S–87	10-in.		19	Nov. 14, 1864	Charleston
S–88	7-in.		21	Feb. 23, 1865	Columbus
S–89	7-in.		25	Jan. 6, 1865	Mobile
S–90	7-in.		28	Dec. 6, 1864	Mobile
S–91	6.4-in.		29		
S–92	6.4-in.	Aug.	2	Dec. 6, 1864	Mobile
S–93	7-in.		4		
S–94	10-in.		6	Dec. 6, 1864	Mobile
S–95	7-in.		8	Jan. 6, 1865	Mobile
S–96	6.4-in.		9	Dec. 21, 1864	Mobile
S–97	11-in.		10	Nov. 14, 1864	Charleston
S–98	7-in.		15	Feb. 23, 1865	Columbus
S–99	7-in.		16	Feb. 23, 1865	Columbus
S–100	7-in.		18	Nov. 14, 1864	Mobile
S–101	10-in.		20	Feb. 24, 1865	Mobile
S–102	6.4-in.		23	March 11, 1865	Columbus
S–103	7-in.	Sept.	1	March 22, 1865	Mobile
S–104	7-in.		2	Dec. 6, 1864	Mobile
S–105	7-in.		7	Dec. 1864	Mobile
S–106	6-pdr.		17		Experimental
S–107	11-in.	Nov.	8	Feb. 18, 1865	Mobile
S–108	11-in.		11	Mar. 3, 1865	Mobile
S–109	6-pdr.		11		Experimental
S–110	10-in.		12		
S–111	7-in.		23	Mar. 17, 1865	Columbus
S–112	11-in.		26	Feb. 18, 1865	Mobile
S–113	11-in.		29	Mar. 11, 1865	Mobile
S–114	7-in.		30		Condemned
S–115	7-in.	Dec.	1		
S–116	11-in.		2		
S–117	7-in.		3		
S–118	30-pdr. Parrott		8		Condemned
S–119	11-in.		9		
S–120	7-in.		13		
S–121	11-in.		14		
S–122	30-pdr. Parrott		17	Feb. 18, 1865	Mobile
S–123	30-pdr. Parrott		17	Mar. 3, 1865	Mobile

NO.	TYPE	CAST	DELIVERED	DESTINATION
S–124	6-pdr.	19		Experimental
S–125	11-in. (Rifle)	19	Mar. 17, 1865	Mobile

24-pounder Coehorn Mortars

NUMBER	CAST	DELIVERED
S–1	July 29, 1864	Condemned
S–2	Aug. 1	Condemned
S–3	1	Condemned
S–2 (sic)	6	Mobile
S–4	10	Mobile
S–5	11	Mobile
S–6	12	Mobile
S–7	13	Mobile
S–8	15	Mobile
S–9	17	Mobile
S–10	17	Mobile
S–11	Sept. 5	Mobile
S–12	7	Mobile
S–13	16	Mobile
S–14	16	Mobile
S–15	23	Mobile
S–16	23	Mobile
S–17	26	Mobile
S–18	26	Mobile

SOURCE: Record of Operations, Selma Naval Ordnance Works, Vols. I and II, National Archives Record Group 45.

Note: S–125 was entered on the last page of Vol. II. It is possible that a Vol. III was begun but misplaced or destroyed. It is known, for example, that no less than twelve 30-pounder Parrot rifles were cast, although only three are entered in the above list. Between August 13 and November 16, 1864, eight 24-pounder Dahlgren boat howitzers were shipped from Selma to Charleston and Mobile. It is uncertain whether or not these were actually cast at Selma.

DIRECTORY OF CONFEDERATE CANNON MANUFACTURERS

ALEXANDER, JOHN & CO.—Columbia, S.C.

The Charleston *Mercury* of March 27, 1862, stated that the churches of Sumter, South Carolina, turned in their bells, about 1500 pounds of metal, to be cast into howitzers by the above.

ANDERSON, J. R. & CO.—Richmond, Va.

Joseph R. Anderson, proprietor. Also known as the Tredegar Iron Works. See Part I.

"ARK. MIL INS."—Arkansas Military Institute

Two Confederate bronze 6-pounders captured January 1, 1864, were inscribed "ARK. MIL. INT." (Albaugh and Simmons, 197) It is known that two cadet 6-pounders were cast for the Arkansas Military Institute by Cyrus Alger of Boston in 1851, one of which survives at Petersburg. (Ripley, 22)

ARMSTRONG & CO.—New Orleans, La.

Col. James Armstrong, proprietor. James Armstrong conducted a large establishment at the corner of Erato and New Levee Sts. which produced all types of machinery. Some projectiles were cast but there is no evidence that cannon were turned out. ("Citizens File") After the war Armstrong associated himself with Mr. John Roy and continued operations at the old location. (New Orleans *Times,* April 22, 1866)

ATLANTA MACHINE WORKS—Atlanta, Ga.

Also known as Gullet & Co. The facility was located on the Georgia Railroad opposite the Richmond & Danville freight depot. Between August and December, 1863, the firm delivered 2323 rounds of shot and shell to the Atlanta Arsenal. (Record of Receipts and Deliveries, Atlanta Arsenal, Chap. 4, Vol. 88, R.G. 109) At that time the facility had seventeen hands of military age in its employ. One source states that cannon were made at this plant and rifled at the roundhouse of the Western & Atlantic Railroad. (Stephen Mitchell, "Atlanta, the Industrial Heart of the Confederacy," *The Atlanta Historical Bulletin* (May, 1930), 23.) *The Memphis Appeal* of November 24, 1861, briefly noted that there was a cannon foundry in Atlanta receiving iron from Dalton, although a specific foundry was not mentioned.

AUGUSTA ARSENAL—Augusta, Ga.

Col. G. W. Rains, commanding. See Part III.

BELLONA FOUNDRY—Chesterfield County, Virginia

Dr. Junius L. Archer, proprietor. See Part I.

BENNETT & LURGES—New Orleans, La.

B.K.T. Bennett and Francis Lurges, proprietors. See Part II.

BRENNAN, T. M. & COMPANY—Nashville, Tenn.

Thomas M. Brennan, proprietor. See Part II.

BRIERFIELD ARSENAL—Selma, Ala.

An 1863 bronze 6-pounder survives at Petersburg which is inscribed "BRIERFIELD ARSENAL." No additional information.

Markings on an 1863 Brierfield Arsenal 6-pounder at the Petersburg National Battlefield Park.

(Tarry Beasley)

BYRAC—New Orleans, La.

The above contracted with the Navy Department for twenty 32-pounder rifles. (*O.N.R.,* Series 1, XVII, 163) No evidence that any were cast.

CAMERON & COMPANY—Charleston, S.C.

Archibald Cameron, proprietor. Cameron & Company cast its first cannon, a large 12-pounder iron rifle, in May, 1861. Work was not completed until August and it was

October before the weapon was delivered to the Charleston Arsenal. Shortly thereafter it was taken to Otter Island for testing where it quickly burst due to poor iron. (Charleston *Mercury,* May 18 and June 17, 1861; Albaugh and Simmons, 207; "Citizens File")

A fire heavily gutted the plant in December, 1861, with damage estimated at $200,000. During the reconstruction interim a temporary foundry was established on King Street and the casting of shot and shell continued. (Charleston *Courier,* Dec. 31, 1861 and Jan. 6, 1862; Charleston *Mercury,* Jan. 28, 1862)

Throughout the war the firm concentrated on the rifling of heavy cannon, the casting of artillery projectiles, and the construction of ship machinery. Few additional cannon were delivered. By March, 1863, two Cameron 6-pounder rifles had been sent to Savannah. (*O.R.*, XIV,690) On July 14, 1863, General Beauregard ordered Brigadier General H. W. Mercer at Savannah: "Send also [to Charleston] two Cammeron [sic] rifled 14-pounder guns, sent from here for siege train. . . ." (*O.R.*, XXVIII, pt. 2, 199) One Cameron 14-pounder rifle survives today in Charleston. (Ripley, 179–180).

CARKEET, J. H.—Natchez, Miss.

Natchez *Courier* of December 11, 1861, reported that the "shop of Mr. J. H. Carkeet on Franklin Street is in progress of manufacturing a new 20-pounder gun invented and patented by him."

CHARLESTON ARSENAL—Charleston, S.C.

See Part III.

CHURCHILL & COMPANY—Natchez, Miss.

The Baton Rouge *Advocate* of May 7, 1861, noted in part that "while strolling through this establishment a few days since, we noticed two of the Natchez brass pieces, one having undergone the process of repair, looking good as new, while the other was revolving in a lathe, taking the first course of a routine mechanical appliance that was to render it, like its companion, more efficient than in its palmiest days." In March, 1862, General Beauregard requested drawings for 6-pounder rifles and 12-pounder Napoleons for "the Natchez foundry." (*O.R.*, pt. 2, 362) No evidence that any were cast. Shortly thereafter the firm was moved to Selma, Alabama, where it continued the production of artillery projectiles.

CLARK, JOHN & COMPANY—New Orleans, La.

John Clark, proprietor. See Part II.

COLUMBUS ARSENAL—Columbus, Ga.

Major F. C. Humphreys, commanding. See Part III.

COLUMBUS IRON WORKS—Columbus, Ga.

W. Riley Brown, proprietor. See Part II.

CONGAREE FOUNDRY—Columbia, S.C.

Two cannon cast by the above are known to survive. No additional information.

CUSHMAN'S FOUNDRY—Houston, Texas

The above produced a smoothbore breechloading cannon invented by George A. Clark. The weapon was loaded with buckshot and was light enough to be carried on a pack saddle for mountain service. (Albaugh and Simmons, 201)

DEANE, F. B. JR. & SON—Lynchburg, Va.

By August, 1861, the Confederate Government had contracted with Deane for forty 12-pounder howitzers. Company papers give no evidence that any were delivered. ("Citizens File") On August 16, 1862, a contract was signed with the Navy Department for 4000 tons of shot and shell. (*O.N.R.*, Series 2, II, 244) Many thousands of projectiles were eventually delivered by the firm. ("Citizens File")

EASON, J. M. & COMPANY—Charleston, S.C.

James M. Eason, proprietor. Soon after the capture of Fort Sumter the Eason firm prepared machinery and began rifling a 24-pounder smoothbore for the State of South Carolina. Subsequently, a portable rifling machine was made which could rifle guns at various distant locations. (———, *The Centennial of Incorporation* (Charleston: The News and Courier Book Press, 1884), 235.)

The Charleston *Mercury* of July 20, 1861, reported that ". . . it has been demonstrated that the Eason gun will throw solid shot or shell, with accuracy, further than any other cannon now in our possession." By September over twenty heavy guns were at the shop awaiting rifling operations. (Columbus *Times,* September 9, 1861) In November, 1862, Major Childs at the Charleston Arsenal reported that "full three weeks have heretofore been taken by Messrs. Eason & Brother in rifling and banding 32-pounder and 42-pounder guns; but that by working at night and on Sundays and distributing the work between Eason and Cameron I hope to be able to finish one gun per week." (*O.R.*, XIV, 690) General Beauregard, on his own initiative and without Richmond's approval, later had several columbiads rifled by Eason. This embroiled him in a controversy with the War Department and initially payment on the bill was refused. Rifling of columbiads had not proven successful in the past but Eason also banded the guns which made them considerably safer. Under growing political pressure, the War Department backed down and the bills were paid. (Ripley, 72–74) Two 10-inch columbiads rifled by Eason survive at Charleston.

ELLIS & MOORE—Nashville, Tenn.

W. D. Ellis and C. C. Moore, proprietors. Engine and boiler making establishment located on South Market St. in Nashville. About sixty hands were employed.

The Nashville *Union and American* of May 5, 1861, and the Nashville *Patriot* of May 29, 1861, stated that the above had cast two sample iron 6-pounders for private parties, one for Louisiana. On May 21, 1861, the State of Tennessee contracted with the firm for a battery of four 6-pounders and two 12-pounder howitzers (all iron) and shot and shell. A payment of $4000 was made at that time. (Military Board Record Book, pp. 21, 25) Company vouchers indicate that no guns were delivered for the Confederate Government. ("Citizens File") A 6-pounder weighing 1010 pounds was captured by Federal at Fort Zollicoffer and marked: "ELLIS & CO. NASHVILLE." (Albaugh and Simmons, 217)

ETOWAH IRON WORKS—Cartersville, Ga.

Mark A. Copper, proprietor. On April 21, 1861, the Confederate Government contracted with the above for both field pieces and 8 and 10-inch columbiads, but Copper proved unable to fill the contract. (Vandiver, 62–63) It was reported that the plant employed 500 hands. (Albaugh and Simmons, 213) The firm was later bought out by William Quinby and William Robinson, formerly of Memphis. There is no indication that any cannon were ever cast at the plant. The War Department at one time studied Cartersville as a possible site for a national cannon foundry, but Selma, Alabama, was later selected. The Columbus *Daily Sun* of May 26, 1864, reported that all the machinery and stores at the plant had been removed by rail and the 250 slaves by wagon train.

GEORGIA RAILROAD MACHINE SHOP—Augusta, Ga.

News article in the Columbus *Daily Enquirer* of August 31, 1861, announced that the mechanics at the above had built a "new styled cannon" called a Sumner Oscillating Breechloading Rifled Gun. It was manufactured from the crank axle of a railroad engine. Other cannon were apparently made at the facility including some 2.25-inch breechloading rifles. (Beverly M. Dubose, III, "The Manufacture of Confederate Ordnance in Georgia," *The Atlanta Historical Bulletin* (1967), 19). There were two breechloading rifles made by the Georgia Railroad Machine Shop and issued through the Atlanta Arsenal on January 28 and April 11. The January 14 issue states however: "Maker: Rushton, Georgia Railroad, Atlanta, Ga." Some 12-pounder Napoleons cast at the nearby Augusta Arsenal were finished at these shops.

GILLELAND, JOHN—Athens, Ga.

Invented a double-barrel experimental smoothbore 4-pounder. The weapon, cast at his Athens foundry, proved a failure. The barrels did not discharge simultaneously and the projectile, two balls linked together by chain, parted and flew in an erratic manner. The weapon was never used beyond the test trial. It survives today at Athens. (Ripley, 180–181)

GOSPORT NAVY YARD—Norfolk, Va.

U.S. Navy Yard occupied by Virginia troops on April 21, 1861, at which time 1198 cannon were captured as follows: one 11-inch columbiad, two 10-inch columbiads, fifty-two 9-inch guns, seventy-nine 8-inch guns, four 64-pounders, 540 32-pounders, a boat and field howitzer (1200 lbs.), two boat and field howitzers (250 lbs.), 235 61 cwt. old style, fifty 70 cwt. old style, forty-four 40 cwt. Shobrick guns, sixty-three 42-pounders 27 cwt. carronades, thirty-five 32-pounders 32 cwt. carronades. ("Report of Ordnance and Ordnance Stores on Hand at the Gosport Navy Yard on the 21st of April, 1861 . . ." Document No. 25, *Documents of the Virginia House of Delaegates, 1861-62* VSL).

By July, 1861, 533 of these guns had been dispersed across the Confedracy as follows: 203—North Carolina, 52—Tennessee, 21—Louisiana, 40—South Carolina, and Georgia, and 217 in 21 batteries in Virginia.

In November, 1861, a Northern informant reported that ". . . the utmost activity prevailed in every department, from the casting of huge Dahlgren guns to the preparation of small shot. (Philadelphia *Inquirer,* November 6, 1861) The probability is, however, that no guns were actually cast under Confederate control, the time being spent in the repair of old cannon.

HA???LESTEN—Memphis, Tenn.

Two Confederate 12-pounder siege guns survive at Fort Donelson. Markings on one are faded but the other, with the exception of three unreadable letters, reads: "MEMPHIS 1862 HA???LESTEN." The 1860 Memphis Directory lists no facility by that name. It is possible that these are some of the 12-pounder siege guns cast by Quinby & Robinson and that "Ha???lesten" was merely an ordnance inspector's name.

HEMPSTEAD, TEXAS

As of June, 1861, a foundry at this location had cast an iron 6-pounder for Confederate troops. (Albaugh and Simmons, 256)

JACKSON, MISSISSIPPI

The authors have seen varied references that some weapons were cast in Jackson, although a specific foundry has never been mentioned. Some Hughes portable breechloading guns, originally made by Street, Hungerford & Company of Memphis, were apparently made in Jackson under order of Governor Pettus, although the exact number and the specific facility are unknown.

JONES' FOUNDRY—Shreveport, La.

The above was under a Confederate contract for both cannon and projectiles. Some projectiles were eventually turned out but no guns.

JONES, McELWAINE & COMPANY—Holly Springs, Miss.

Founded in 1859 and initially involved in the manufacture of iron grill work sold mainly in New Orleans. Some 200 hands were employed at the beginning of the war. The firm was primarily involved in the repair of small arms. A few sample bronze cannon were also made. McElwaine's widow later recalled how she used to assist in pouring out the ladles of metal into the mold.

LEEDS & COMPANY—New Orleans, La.

Charles J. and Thomas L. Leeds, proprietors. See Part II.

LIVERMORE FOUNDRY & MACHINE WORKS—Memphis, Tenn.

Reported to have made some cannon for the C. S. until Memphis fell and thereafter made cannon for the U.S. (Albaugh and Simmons, 240)

MACON ARSENAL—Macon, Ga.

Maj. Richard M. Cuyler, commanding. See Part III.

MILLER, A. N.—Savannah, Ga.

When Northern public sentiment caused the Fort Pitt Foundry of Pittsburgh to decline delivery of contracted cannon to Georgia in 1860, the State not only contracted with the Tredegar Iron Works but also turned to her own resources. In December, 1860, the A.N. Miller foundry made a proposal to furnish six 32-pounder seacoast guns "of the most modern pattern United States standard. . . ." The proposal was later changed to 24-pounders but to no avail, since the firm proved unable to deliver heavy caliber guns at that time. (Miller to Brown, Dec. 12 and Dec. 21, 1860, Georgia State Archives) Further effort at encouraging heavy cannon production, including a $10,000 State contract, proved fruitless. (*O.R.*, Series 4, I, 169–170)

The State later contracted with Miller for a battery of 24-pounder howitzers. The first weapon, made of iron and weighing 1700 pounds, was delivered in May 1861. At that time the firm had an order from South Carolina for 6000 rounds of shot and shell and a standing order for an unlimited quantity of the same for Georgia. (Albaugh and Simmons, 246)

By December, 1861, the firm was engaged in "the casting of guns and mortars of large calibre and superior workmanship." A 32-pounder was taken from the shop on December 8th and mounted in the Savannah defenses. (Columbus *Daily Times*, December 9, 1861) Additional production was limited to one 15-inch and three 10-inch seacoast mortars, all cast in 1862 and placed in the Savannah defenses. Some old heavy guns were also rifled and banded. ("Citizens File"; *O.R.*, XIV, 874)

NOBLE BROTHERS & COMPANY—Rome, Ga.

James Noble, Sr., proprietor. See Part II.

PALMETTO IRON WORKS—Columbia, S.C.

Founded in 1850 by William Glaze and located at the corner of Laurel and Lincoln Streets. Prior to the war the firm entered into a contract with the State of South Carolina for small arms and swords. During the war several revolving cannon were made. The weapons were invented by a man named "George" and they worked on the order of Colt's revolvers. The guns were never adopted by the Government. (Albaugh and Simmons, 252)

*PAXTON, A. M. & COMPA*NY—Vicksburg, Mississippi

The above firm did finishing work on some of the field guns cast by Quinby & Robinson and A. B. Reading & Brother. There is no evidence that the firm actually cast guns of its own.

PHOENIX IRON WORKS—Gretna (New Orleans), La.

Sylvester Bennett, proprietor. News item dated May 5, 1861, noted that the above had cast a heavy gun eight feet in length. The New Orleans *Picayune* of September 4, 1861, noted that several 18-pounder cannon had been turned out at the firm.

QUINBY & ROBINSON—Memphis, Tenn.

William T. Quinby and William A. Robinson, proprietors. See Part II.

READING, A. B. & BROTHER—Vicksburg, Miss.

Abram B. Reading, proprietor. See Part II.

RICHMOND NAVAL ORDNANCE WORKS—Richmond, Va.

The above facility rifled, banded, and bored some of the heavy guns cast at the nearby Tredegar Iron Works after that firm was struck by fire in May, 1863. (*O.N.R.*, Series 2, II, 548, 641, 757) There is a surviving Tredegar 6.4-inch Brooke rifle at the Washington Navy Yard marked "RNOW" (Richmond Naval Ordnance Works), indicating that work was performed on the gun at the facility. Also inscribed on this and several other surviving guns are the initials "A.D.B.", which stands for Lieutenant A. M. deBee who was commander of the works as of October, 1863. By February, 1865, the Richmond Naval Ordnance Works employed 115 whites and forty-two Negroes. (Misc. Letters, Ordnance and Hydrography Department, National Archives Record Group 45)

SAMSON & PAE—Richmond, Va.

Thomas Samson and James Pae, proprietors. This Richmond machine shop did finishing work on the howitzers cast by the Washington Foundry of Richmond. The firm also made some weapons of its own. On April 23, 1862, a wrought iron cannon (caliber unspecified) was delivered to the Confederate Government. On January 16, 1863, four Williams breechloading guns were delivered at $900 apiece. The facility was also a large supplier of artillery projectiles. ("Citizens File")

SELMA NAVAL GUN FOUNDRY—Selma, Ala.

Catesby ap R. Jones, commanding. See Part III.

SHACKLOCK & COMPANY—Memphis, Tenn.

A cannon sight base survives which is marked "SHACKLOCK & CO., MEMPHIS, TENN." According to the 1860 Memphis Directory W. D. Shacklock was a Memphis jeweler. During the war Shacklock turned his skills to the manufacture of fine artillery implements and the sighting of cannon. The Shacklock firm sighted scores of cannon at Columbus, Kentucky, and at fortifications along the Mississippi River. Records indicate that the firm actually cast no cannon of its own. ("Citizens File")

SKATES & COMPANY—Mobile, Ala.

Skates & Company, also known as the Mobile Foundry, was located at the corner of Magnolia and State Streets. Prior to the war the firm manufactured steam engines, boilers, gin gearings, and iron machinery.

During the war the Skates firm engaged primarily in the rifling and banding of heavy guns and the casting of artillery projectiles. (Memphis *Appeal*, September 28,

1861; "Citizens File") In September, 1861, a contract was signed with the firm for four field batteries complete with harness. (*O.R.*, VI. 726, 729) Company vouchers give evidence that only two cannon were delivered—a rifled bronze gun caliber 3.67 with carriage complete delivered October 29, 1861, at $1000 and a bronze 12-pounder howitzer with carriage complete delivered December 14, 1861, at $900. ("Citizens File")

The Charleston *Mercury* of April 2, 1862, reported: "The Government has copper enough in Mobile to make four or five brass cannon, but it cannot for the want of tin, which cannot be had, or at least is very scarce . . . The foundries at Mobile can make no brass cannon at present. Their whole time is devoted to making more useful articles namely iron cannon and shot and shells and rifling cannon."

On March 21, 1862, Skates & Company addressed Colonel H. Oladowski of the Ordnance Department that "we have completed the iron Parrott guns and disposed of them to the State—Ala. We have done some excellent shooting with 8 ounces of powder with a range of 2¾ miles at all the elevation we could get." ("Citizens File") It is uncertain whether or not this was a reference to guns actually having been cast at the foundry.

STATE MILITARY WORKS—Greenville, S.C.

The State Military Works was put in operation in late 1862 and by early 1863 employed 75 hands in the manufacture of breechloading small arms. The manufacture of cannon and gun carriages was also contemplated at that time. A specimen of iron ore from Kings Mountain was sent to Richmond to ascertain its strength for ordnance use. The ore was reported to have been cast into a cannon. (Albaugh and Simmons, 265) The Tredegar Iron Works did cast a 24-pounder rifle and a 12-pounder howitzer from South Carolina iron on April 21, 1862, but whether or not this was the same gun cannot be certain. The railroad bridges between Greenville and Columbia were strengthened to permit the passage of trains with cannon. Whether or not any cannon were eventually cast in Greenville is unknown.

STREET, HUNGERFORD & COMPANY—Memphis, Tenn.

Anthony S. Street and Fayette H. Hungerford, proprietors. See Part II.

TAPPY & LUMSDEN—Petersburg, Va.

The Petersburg *Express* of May 11, 1861, noted that the above had completed a five-shot cannon of 3.13-inch caliber weighing 300 pounds. It fired a four pound ball 1900 yards. A pair of these weapons was turned out. One burst during firing tests and the other was set aside for the duration of the war. It survives today at Petersburg. (Ripley, 181–182) The firm later turned out battery equipment. Between October, 1861, and July, 1863, deliveries were made of fifty-two gun carriages twenty-nine field and twenty-three heavy), thirty-two caissons, and one battery wagon. ("Citizens File")

TREDEGAR IRON WORKS—Richmond, Va.

See J. R. Anderson & Co.

WASHINGTON FOUNDRY—Richmond, Va.

William J. Hubard, proprietor. See Part II.

WEBSTER, THOMAS & COMPANY—Chattanooga, Tenn.

The State of Tennessee contracted with the above on June 17, 1861, for a battery of four 6-pounder and two 12-pounder howitzers with carriages, caissons, shot and shell. At that time the firm was advanced $4000. (Military Board Record Book, pp. 39, 45) A report to the Tennessee General Assembly made October 1, 1861, stated that "cannon enough have been cast, both iron and bronze, to supply the whole Provisional Army of Tennessee for the present. This has been done principally in Nashville and Memphis, and to some extent in Chattanooga. . . ." (*O.R.*, LII, pt. 2, 162–163)

WHITFIELD, BRADLEY & COMPANY—Clarksville, Tenn.

Besides doing finishing work on some guns cast by the T. M. Brennan foundry, there is evidence that the above cast at least four guns (two iron 6-pounders and two bronze 9-pounders) of its own. The weapons served with Tilghman at Hopkinsville, Kentucky, and later at Fort Donelson. Tilghman described the guns as "pretty fair" but another officer remarked that the 9-pounders were "of very little account". (*O.R.*, IV, 464, 514; VII, 388, 394)

WOLFF, S. & COMPANY—New Orleans, La.

Samuel Wolff, proprietor. See Part II.

WRIGHT & RICE—Florence, Ala.

By August, 1861, the Confederate Government had undertaken a contract with Wright & Rice for forty 24-pounder siege howitzers with carriages and caissons. (Lee, 167–168) On November 4, 1861, John B. Read wrote Brigadier General Withers at this Mobile headquarters concerning the firm. He wrote in part: "Mr. Wright, the ranking shop partner, is very skillful in his line. The firm will take contracts for guns of any size for [which] they want at the average rate paid for such work by the government. They use a blast furnance; but so does the foundry at Huntsville from which Col. Gorgas has ordered a number of guns for Richmond. They use Mr. Kirkman's iron from [the] Cumberland river, and Mr. Wright has great confidence in it . . . They have patterns now ready for casting eight inch Siege Howitzers if you should want any of that size. They have also drawings for 24-pdr. Siege Howitzers sent them by Col. Gorgas. They have no seasoned timber for gun carriages. None of the desired metals are to be had in Florence." ("Citizens File") The firm did complete a breechloading cannon which was reported to fire ten times a minute. (Albaugh and Simmons, 277) There is no evidence that any additional weapons were produced.

YOUNG, J. R. & COMPANY—Huntsville, Ala.
J. R. Young, proprietor. See Part II.

ZARRIMAN & HILZEN—Athens, Tenn.

The above firm declined a May 22, 1861, proposal from the State of Tennessee to cast cannon. (Military Board Record Book, 22)

THE SCHOFIELD IRON WORKS, MACON, GA.

The above photograph was typical of so many Southern foundries that converted to a war production. The Schofield foundry actually cast no cannon, but a number of projectiles and some powder machinery were turned out. *(Herb Peck)*

Inside the Schofield Iron Works (*Herb Peck*)

APPENDIX

ESTIMATION OF CANNON PRODUCTION IN MAJOR FOUNDRIES
April 1861–April 1865

FOUNDRY	LOCATION	FIELD	HEAVY	TOTAL
	C.S.A.			
Tredegar Iron Works	Richmond, Va.	612	431	1043
Naval Gun Foundry	Selma, Ala.	20	133	153
Bellona Foundry	Chesterfield Co., Va.		135	135
Augusta Arsenal	Augusta, Ga.	130		130
Quinby & Robinson	Memphis, Tenn.	95	5	100
John Clark & Co.	New Orleans, La.	100		100
Noble Bros. & Co.	Rome, Ga.	75	10	85
T. M. Brennan & Co.	Nashville, Tenn.	76	7	83
Columbus Arsenal	Columbus, Ga.	80		80
Macon Arsenal	Macon, Ga.	73	1	74
Leeds & Co.	New Orleans, La.	55	10	65
A. B. Reading & Bro.	Vicksburg, Miss.	45		45
	U.S.A.			
West Point Foundry	Cold Spring, N.Y.	806	739	1545
Fort Pitt Foundry	Pittsburgh, Pa.		995	995
Phoenix Iron Co.	Phoenixville, Pa.	857		857
Cyrus Alger & Co.	Boston, Mass.	387	311	689
Ames Mfg. Co.	Chicopee, Mass.	554		554
Revere Copper & Co.	Boston, Mass.	461		461
Henry N. Hooper & Co.	Boston, Mass.	376		376
Seyfort, McManus & Co.	Reading, Pa.	100	127	227
M. Greenwood & Co.	Cincinatti, Ohio	151		151
O(?) Foundry	New York, N.Y.	60		60
Builders Foundry	Providence, R.I.		50	50
Hinkley, Williams & Co.	Boston, Mass.		50	50
Marshall & Co.	St. Louis, Mo.	45		45

SOURCES: The figures for the Tredegar Iron Works are based on the Tredegar Gun Foundry Book, Virginia State Library. The Selma Naval Gun Foundry totals are from the "Selma Foundry Record of Operations," National Archives. The 20 field guns cast at Selma were experimental 6-pounders, all of which burst during firing tests. The heavy gun total for that facility includes 19 coehorn mortars. No records are available for the Bellona Foundry, Augusta Arsenal, and Columbus Arsenal and the estimated figures are based upon the foundry numbers of surviving weapons. The Macon Arsenal total is based on the "Macon Arsenal Daily Reports of Ordnance," National Archives. The remaining private firm figures are based primarily upon company vouchers in the "Citizens File" and all other available sources, including newspaper accounts, the *Official Army and Naval Records,* and State Archival holdings.

The Northern estimations have a much higher degree of accuracy due to availability of records. Most figures are based on the information in "U. S. House of Representatives, Document No. 99, 2nd Session, 40th Congress, 1867–68," National Archives. Some approximations were involved, however. The deliveries of the Ames Manufacturing Company, for example, did not include any James 14-pounder rifles, a gun made exclusively by that firm. The figure 100, the estimated number of James guns cast, is therefore added to the total. The precise New York City Foundry (O? Foundry) that cast Wiard guns is unknown. The figure represents an estimation.

TABLE OF CANNON CAST AT J. R. ANDERSON & COMPANY

NOTE

The vast majority of information recorded in the following Table is taken verbatim as it was transcribed in the Tredegar Gun Foundry Book, Virginia State Library. It must be remembered that the clerks were not writing history. It is very possible that entries were not made daily but perhaps weekly. This would account for the fact that foundry numbers on some surviving weapons on occasion do not match the Table. A bronze 12-pounder howitzer, foundry number 1400, survives at the Shiloh park. No. 1400 was recorded as an "8-in. How." This discrepency does not mean that an 8-inch Siege Howitzer was not cast, but simply attests to poor bookkeeping.

Other discrepancies complicate the Gun Foundry Book. On a number of occasions entries were made simply as "12-pdr.," giving no indication whether this was a reference to a 12-pounder howitzer, 12-pounder Napoleon, or 12-pounder siege gun. 6-pounder iron smoothbores and 3-inch rifles were grouped together in the Gun Foundry Book as simply "6-pdr. Iron." In some instances no indication was given as to whether or not a 6-pounder or 12-pounder howitzer was an iron or bronze model.

It soon became evident that to transcribe the Gun Foundry Book verbatim would create so much confusion as to render the Table useless at points. The authors, therefore, decided to compile a Table "with interpretation." To definitely establish the identity of certain guns it became necessary to search additional sources. The Tredegar Foundry Sales Book, Virginia State Library, and J. R. Anderson & Company, "Citizens File" were used. Also, valuable information on surviving Tredegar weapons was provided by Mr. Edwin Olmstead. Our thanks is extended to Mr. Olmstead for the numerous letters exchanged on this subject.

In most cases these additional sources clarified specifically what type of weapon and what metal the entry was a reference to. Relating to "6-pdr. irons," thirty were found to be 6-pounder smoothbores and fourteen were 3-inch rifles. Unfortunately, thirty-one "6-pdr. irons" remain a mystery as to specific identification.

Difficulties were also encountered with Parrott rifles. A number of early entries were entered in the Gun Foundry Book as "6-pdr. Parrotts." It was originally believed by the authors that this was a reference to banded 6-pounder rifles. Additional search and correspondence on this matter, however, has found that this was simply an early name for 10-pounder Parrotts. The "6-pdr. Parrott" term has been retained in the Table, but the reader is to understand this to be a reference to 10-pounder Parrotts. Some entries were made as "Long Tom." This was a reference to 30-pounder Parrotts. A Union 30-pounder was captured at First Manassas and dubbed "Long Tom." Tredegar based its model on this gun. The term "Long Tom" has not been used in the Table. All entries are listed as "30-pdr. Parrotts."

In the Gun Foundry Book some entries are listed as simply "4.62" and "7-in. Rifle Army." Because definite identification cannot be established, these entries are not changed. It is believed by the authors, however, that these are references to Brooke guns. Again, an apparent example of poor bookkeeping.

In most cases information has been found to identify whether or not a 6-pounder or 12-pounder howitzer was iron or bronze. The words "brass" and "bronze" were used interchangeably. All 12-pounder Napoleons cast in 1864–65 were of iron.

On rare occasions the authors admit to taking liberties where records do not specifically establish an identity. Numbers 2249, 2250, 2251, and 2256, for example, were listed originally as "12-pdr." Likewise, numbers 2254, 2255, and 2257 are listed as "10-pdr." On the assumption that this is another example of poor bookkeeping, the Table has these listed as Napoleons and Parrotts respectively.

The reader must remember that the date recorded was the "date of casting." It usually took a minimum of one month and much longer in some cases to prepare a gun for the field. Dates and foundry numbers are occasionally not in consecutive order in the Gun Foundry Book. This has been left as is in the Table. Some foundry numbers were skipped altogether. Whether or not these guns were cast and simply not recorded is not known. Only the bore diameters were recorded for most of the heavy guns cast in 1861. More specific information was usually recorded after that date. Foundry numbers 1164–1213 were recorded on a page in the Gun Foundry Book which was later half destroyed. About thirty heavy guns and five field pieces (including three surviving bronze 3-inch rifles) were cast during this period. Only the weights of the various guns remains and not the specific calibers.

The following abbreviations are used in the Table:

in.	= inch	How.	= Howitzer
pdr.	= pounder	S.B.	= Smoothbore

1861

April	3	1158	10-in.
	4	1160	8-in.
	5	1159	10-in.
	6	1161	7-in.
	9	1163	10-in.
(See Beginning Notes)			
May	22	1214	10-in.
		1215	9-in.
	25	1216	7-in.
	27	1217	7-in.
		1218	24-pdr. How. Iron
		1219	24-pdr. How. Iron
		1220	24-pdr. How. Iron
		1221	24-pdr. How. Iron
	28	1222	10-in.
	30	1224	9-in.
	31	1225	10-in.
June	1	1227	8-in.
	4	1226	10-in.
	6	1229	8-in.
	8	1230	10-in.
	14	1231	10-in.
	17	1233	8-in.
	18	1232	10-in.
	20	1234	8-in.
	25	1235	10-in.
	26	1236	8-in. Rifle
	27	1237	8-in.
	28	1238	10-in.
	29	1239	10-in.
July	1	1240	10-in.
	3	1241	8-in.
	8	1245	9-in. Rifle
	9	1242	10-in.
	12	1240	10-in.
		1244	10-in.
	18	1247	8-in.
	20	1246	10-in.
	23	1248	8-in.
	25	1249	8-in.
	29	1251	10-in.
	30	1250	8-in.
	31	1252	10-in.
Aug.	2	1253	8-in.
	3	1254	8-in.
	6	1255	10-in.
	8	1256	8-in. How.
		1266	8-in. How.
	10	1267	10-in.
	12	1268	8-in.
	13	1269	8-in. How.
		1270	8-in. How.
	16	1271	10-in.
Sept.	12	1272	8-in.
	16	1273	8-in.
	18	1274	8-in.
	19	1275	10-in.
	21	1276	10-in.
	25	1277	32-pdr.
	27	1278	32-pdr.
	29	1279	8-in. How. Rifle
	30	1280	8-in. How. Rifle
Oct.	2	1281	8-in.
	4	1282	10-in.
	12	1283	8-in.
	14	1284	8-in.
	17	1285	10-in.
	19	1286	10-in.
	21	1287	9-in. Navy
	23	1288	8-in.
	24	1289	10-in.
	26	1290	8-in.
	29	1291	8-in.
	31	1292	8-in.
Nov.	2	1293	9-in.
		1294	8-in.
	4	1295	10-in. Rifle
	5	1296	8-in.
	6	1297	10-in.
	8	1298	10-in.
	9	1299	8-in.
	11	1300	8-in. Rifle
	13	1301	9-in.
		1302	12-pdr. How. Iron
	14	1304	6-pdr. Iron
		1303	8-in.
		1305	6-pdr. Iron
	16	1306	10-in.
	18	1307	8-in.
		1308	6-pdr. Parrott Iron
		1309	6-pdr. Parrott Iron
	22	1310	32-pdr. Rifle
		1311	12-pdr. How. Iron
	23	1312	9-in. Navy Rifle
		1313	6-pdr. Parrott with loop for Navy
		1313	6-pdr. Parrott with loop for Navy
	25	1315	32-pdr. Rifle
	26	1316	10-in.
		1317	6-pdr. Parrott Iron
	27	1318	6-pdr. Parrott Iron
	28	1319	8-in.
	29	1320	9-in. Navy Rifle

Dec.	2	1321	8-in.
	4	1322	10-in.
		1323	6-pdr. Parrott Iron
	5	1324	8-in.
		1325	12-pdr. How. Iron
	7	1326	9-in. Rifle
		1327	6-pdr. Parrott Iron
		1328	12-pdr. How. Brass
	9	1329	8-in.
	10	1330	12-pdr. How. Brass
	11	1331	10-in.
	13	1332	12-pdr. How. Brass
		1333	24-pdr. Rifle
		1334	6-pdr. Rifle Iron
		1335	12-pdr. How. Brass
	17	1336	6-pdr. Iron
		1337	8-in.
	19	1338	10-in.
		1339	12-pdr. How. Brass
		1341	24-pdr. Rifle
		1342	8-in. How.
		1343	6-pdr. Iron
		1340	6-pdr. Iron
		1344	32-pdr. Rifle
		1345	Mountain Rifle
		1346	Mountain Rifle
		1347	8-in. How.
		1348	24-pdr. Rifle
	30	1349	6-pdr. Rifle Iron
		1350	6-pdr. Rifle Iron
			1862
Jan.	2	1351	32-pdr. Rifle
		1352	6-pdr. Iron
		1353	6-pdr. Iron
	5	1354	8-in. How.
	4	1355	32-pdr. Rifle
	7	1356	32-pdr. S.B.
		1357	24-pdr. Rifle
		1358	12-pdr. How. Iron
		1359	12-pdr. How. Iron
		1360	12-pdr. How. Brass
		1431	12-pdr. How. Iron
		1432	12-pdr. How. Iron
	10	1361	8-in. How.
	9	1362	32-pdr. Rifle
	10	1363	8-in. How.
	13	1364	10-in.
	14	1365	6-pdr. Iron
		1366	6-pdr. Iron
		1367	32-pdr.
	15	1368	8-in. How.
		1369	8-in. How.
	17	1370	32-pdr.
	16	1371	12-pdr. How. Brass

		1372	Mountain Rifle, Brass
		1373	3-in. Rifle
		1374	3-in. Rifle
	27	1375	9-in.
		1376	6-pdr. Parrott
		1379	24-pdr. Rifle
	31	1380	32-pdr. Rifle
Feb.	3	1381	24-pdr. Rifle
	2	1382	6-pdr. Parrott
		1383	6-pdr. Parrott
	4	1385	9-in.
	5	1386	12-pdr. Mountain How. Brass
		1387	12-pdr. Mountain How. Brass
	7	1388	6-pdr. Rifle Iron
		1389	9-in.
	10	1390	12-pdr. Mountain How. Brass
		1391	12-pdr. Mountain How. Brass
	11	1392	6-pdr. Rifle Iron
	12	1393	12-pdr. Mountain How.
		1394	12-pdr. Mountain How.
	14	1395	32-pdr. Navy
	15	1396	24-pdr. Rifle
	21	1397	9-in. Navy
	24	1398	24-pdr. Rifle
	26	1399	8-in. How.
		1400	8-in. How.
	27	1401	8-in. Columbiad
March	1	1402	24-pdr. Rifle
	3	1403	8-in.
	4	1404	10-in.
		1405	8-in.
	8	1406	10-in.
		1407	32-pdr.
		1408	6-pdr. Iron
		1409	6-pdr. Iron
		1410	6-pdr. Iron
		1436	6-pdr. Iron
		1437	6-pdr. Iron
		1438	6-pdr. Iron
		1439	6-pdr. Iron
		1440	6-pdr. Iron S.B.
		1441	6-pdr. Iron S.B.
		1442	24-pdr. How. Iron
		1435	12-pdr. How. Brass
		1443	6-pdr. Brass
		1444	6-pdr. Brass
		1445	6-pdr. Brass
		1446	6-pdr. Brass
		1447	12-pdr. How. Brass
		1448	12-pdr. How. Brass
	29	1449	12-pdr. How. Iron
		1450	6-pdr. Iron S.B.
		1451	6-pdr. Iron S.B.
		1452	6-pdr. Iron S.B.
	28	1453	3-in. Rifle

		1454	3-in. Rifle
		1455	3-in. Rifle
		1456	3-in. Rifle
		1457	24-pdr. How.
		1458	12-pdr. How. Brass
Apr.	4	1459	12-pdr. How. Brass
Mar.	31	1460	6-pdr. Iron S.B.
		1461	3-in. Rifle
		1462	3-in. Rifle
		1463	3-in. Rifle
		1464	3-in. Rifle
Apr.	2	1465	6-pdr. Iron S.B.
		1466	6-pdr. Iron S.B.
		1467	6-pdr. Iron S.B.
		1468	12-pdr. How. Iron
		1469	12-pdr. How. Iron
	4	1470	24-pdr. How. Iron
		1471	6-pdr. Iron S.B.
		1472	6-pdr. Iron S.B.
		1473	12-pdr. How. Iron
	7	1474	12-pdr. How. Iron
		1475	6-pdr. Iron S.B.
		1476	6-pdr. Iron S.B.
		1477	6-pdr. Iron S.B.
	8	1478	3-in. Rifle
		1479	3-in. Rifle
		1480	3-in. Rifle
		1481	12-pdr. How. Iron
	14	1482	10-in.
	9	1483	24-pdr. How. Iron
		1484	12-pdr. How. Iron
		1485	6-pdr. Iron S.B.
		1486	6-pdr. Iron S.B.
		1487	6-pdr. Iron S.B.
	10	1488	6-pdr. Iron S.B.
		1489	6-pdr. Iron S.B.
		1490	6-pdr. Iron S.B.
		1491	6-pdr. Iron S.B.
		1433	Mountain Rifle Brass
		1434	Mountain Rifle Brass
	11	1492	12-pdr. How. Iron
		1494	6-pdr. Iron S.B.
		1495	6-pdr. Iron S.B.
		1496	3-in. Rifle
	14	1497	12-pdr. How. Iron
		1498	3-in. Rifle
		1499	3-in. Rifle
		1500	6-pdr. Iron S.B.
		1501	6-pdr. Iron S.B.
	15	1502	12-pdr. How. Iron
		1503	6-pdr. Iron S.B.
		1504	6-pdr. Iron S.B.
		1505	24-pdr. How. Iron
	16	1506	12-pdr. How. Iron
		1507	12-pdr. How. Iron
		1508	3-in. Rifle
		1509	6-pdr. Iron S.B.
	17	1510	3-in. Rifle
		1512	12-pdr. How. Iron
		1513	12-pdr. How. Iron
		1511	6-pdr. Iron S.B.
		1514	24-pdr. Rifle
		1515	Mountain Rifle Brass
		1516	Mountain Rifle Brass
	18	1517	12-pdr. How. Iron
		1518	6-pdr. Iron S.B.
		1519	6-pdr. Iron S.B.
	19	1520	6-pdr. Iron S.B.
		1521	6-pdr. Iron S.B.
		1522	6-pdr. Iron S.B.
		1523	12-pdr. How. Iron
		1524	12-pdr. How. Iron
	21	1525	24-pdr. Rifle S.C. Iron
		1526	12-pdr. How. S.C. Iron
		1527	6-pdr. Iron
		1528	6-pdr. Iron
	24	1529	12-pdr. How. Brass
		1530	12-pdr. How. Brass
	28	1531	6-pdr. Brass Eddenton Bells
	29	1532	12-pdr. How. Eddenton Bells
		1533	12-pdr. How. Eddenton Bells
		1534	6-pdr. How. Eddenton Bells
May	1	1535	6-pdr. Brass
		1536	12-pdr. How. Brass
	2	1537	6-pdr. Brass
		1538	6-pdr. Brass
	3	1539	12-pdr. How. Brass
		1540	6-pdr. Brass
	4	1541	6-pdr. Brass
		1542	6-pdr. Brass
		1543	12-pdr. How. Brass
	6	1544	6-pdr. Bronze
		1545	6-pdr. Bronze
	7	1546	12-pdr. How. From Bell Metal
	8	1547	12-pdr. How. From Bell Metal
		1548	12-pdr. How. From Bell Metal
		1549	12-pdr. How. From Bell Metal
		1550	Mountain Rifle Brass
		1551	Mountain Rifle Brass
		1552	Mountain Rifle Brass
	12	1553	6-pdr. Brass
	13	1554	12-pdr. Mountain How.
		1555	12-pdr. Mountain How.
		1556	12-pdr. How. Bronze
	14	1557	12-pdr. How. Bronze
		1558	12-pdr. Mountain How.
		1559	12-pdr. Mountain How.
	15	1560	12-pdr. How. Brass
	20	1561	12-pdr. How. Iron
		1562	12-pdr. How. Iron

		1563	3-in. Rifle
		1564	3-in. Rifle
	23	1565	7-in. Rifle Brooke Pattern
	24	1566	12-pdr. Mountain How. Brass
		1567	12-pdr. Mountain How. Brass
	26	1568	12-pdr. How. Brass
		1569	12-pdr. How. Brass
	27	1570	Mountain Rifle Brass
		1571	Mountain Rifle Brass
		1572	32-pdr. Rifle Navy
	30	1573	12-pdr. How. Brass
		1574	Mountain Rifle
		1575	Mountain Rifle
	31	1576	7-in. Brooke Rifle
		1577	12-pdr. How. Brass
		1578	12-pdr. How. Brass
June	5	1579	Mountain Rifle
		1580	Mountain Rifle
	4	1581	6-pdr. Brass Bad
		1582	6-pdr. Brass Bad
	6	1583	7-in. Brooke Rifle
		1584	12-pdr. Mountain How.
		1585	12-pdr. Mountain How.
		1586	12-pdr. How. Brass
	9	1587	32-pdr. Rifle Brooke
		1588	6-pdr. Brass Bad
		1589	6-pdr. Brass Bad
		1590	6-pdr. Parrott Banded
		1591	6-pdr. Parrott Banded
	10	1592	6-pdr. Parrott Banded
		1593	6-pdr. Parrott Banded
		1594	6-pdr. Iron S.B.
		1595	12-pdr. How. Iron
	11	1596	6-pdr. Brass
	13	1597	7-in. Rifle Brooke Triple Banded
	14	1598	6-pdr. Brass
		1599	6-pdr. Brass
		1600	6-pdr. Iron S.B.
		1601	6-pdr. Iron S.B.
		1602	12-pdr. How. Iron
		1603	12-pdr. How. Iron
		1604	12-pdr. How. Brass
		1605	10-in. Columbiad
		1606	6-pdr. Iron
		1607	6-pdr. Iron
		1608	6-pdr. Iron
	19	1609	6-pdr. Iron
		1610	32-pdr. Brooke
	20	1611	7-in. Brooke Rifle
		1612	12-pdr. How. From James Rifle
		1613	Mountain Rifle Brass
		1614	Mountain Rifle Brass
	26	1615	10-in. Columbiad
		1616	12-pdr. Mountain How.
		1617	12-pdr. How. Bronze
		1618	12-pdr. How. Bronze
		1619	12-pdr. How. Bronze
		1620	6-pdr. Bronze
		1621	10-in. Columbiad
July	3	1622	32-pdr. Rifle
	7	1623	10-in. Mortar
	9	1624	10-in. Columbiad
	10	1625	24-pdr. Rifle
	15	1626	24-pdr. Rifle
	21	1627	24-pdr. Rifle
		1628	6-pdr. Brass
		1629	6-pdr. Brass
	26	1630	7-in. Brooke Rifle
		1631	30-pdr. Parrott
		1632	12-pdr. Banded New Pattern
		1633	32-pdr. Brooke
		1634	10-in. Mortar
		1635	12-pdr. How. Brass
		1636	9-in. Navy
		1637	30-pdr. Parrott
		1638	12-pdr. Banded Rifle New Pattern
		1639	12-pdr. Mountain How. Brass
		1640	10-in. Mortar
Aug.	8	1641	7-in. Brooke Rifle
	11	1642	10-in. Columbiad
		1643	12-pdr. Banded New Pattern
		1644	12-pdr. Napoleon Brass
	13	1645	30-pdr. Parrott
		1646	10-in. Mortar
	14	1647	32-pdr. Brooke
	16	1648	10-in. Columbiad
	19	1649	30-pdr. Parrott Lost
		1650	10-pdr. Parrott
	20	1651	10-in. Mortar
		1652	7-in. Brooke Rifle
		1653	10-in. Mortar
		1654	6-pdr. Parrott
		1655	6-pdr. Parrott
		1656	10-in. Columbiad
		1657	10-in. Mortar
		1658	20-pdr. Parrott
	27	1659	10-in. Columbiad
		1660	10-in. Mortar
		1661	20-pdr. Parrott
		1662	30-pdr. Parrott
		1663	32-pdr. Brooke Rifle
Sept.	1	1664	10-in. Columbiad
		1665	10-in. Columbiad
		1666	10-in. Mortar
		1667	10-in. Mortar
		1670	12-pdr. Napoleon Brass
		1668	10-in. Columbiad
Oct.	5	1671	32-pdr. Brooke
	7	1672	10-in. Columbiad
	11	1673	30-pdr. Parrott

		1674	10-pdr. Parrott
		1675	10-pdr. Parrott
	13	1676	10-in. Columbiad
		1678	10-in. Columbiad
Oct.	24	1679	7-in. Brooke Army
	27	1680	7-in. Brooke Army
	30	1681	10-in. Columbiad
Nov.	3	1682	7-in. Brooke
		1683	12-pdr. Napoleon Brass
		1684	12-pdr. Napoleon Brass
		1685	12-pdr. How. Brass
		1686	12-pdr. How. Brass
	8	1687	10-in. Columbiad
	10	1689	10-in. Columbiad
	12	1690	30-pdr. Parrott
		1691	10-pdr. Parrott
		1692	7-in. Brooke
		1693	12-pdr. Napoleon Brass
		1694	12-pdr. Napoleon Brass
		1695	12-pdr. Mountain How.
	21	1696	10-in. Columbiad
	24	1697	7-in. Brooke
	25	1698	30-pdr. Parrott
		1699	20-pdr. Parrott
		1700	10-pdr. Parrott
		1701	20-pdr. Parrott
		1702	20-pdr. Parrott
	29	1703	32-pdr. Parrott
		1704	20-pdr. Parrott
		1705	10-pdr. Parrott
Dec.	1	1706	10-in. Columbiad
	3	1707	32-pdr. Brooke
	4	1708	12-pdr. Napoleon Brass
	6	1709	7-in. Brooke Long Gun
		1710	20-pdr. Parrott Gorgas Pattern
		1711	32-pdr. Brooke
		1712	20-pdr. Parrott
	8	1713	10-in. Columbiad
	10	1714	32-pdr. Brooke
		1715	12-pdr. Napoleon Brass
	12	1716	12-pdr. Napoleon Brass
	13	1717	7-in. Brooke Long Gun
	16	1718	32-pdr. Brooke
	17	1719	10-in. Brooke
	19	1720	4.62-in. Rifle
		1721	10-pdr. Parrott
	22	1722	7-in. Rifle Army
		1723	12-pdr. Napoleon Brass
	24	1724	10-in. Columbiad
	27	1725	4.62-in. Rifle
		1726	20-pdr. Parrott
	29	1727	7-in. Brooke Rifle Navy
	31	1728	32-pdr. Brooke Rifle
		1729	20-pdr. Parrott

1863

Jan.	5	1730	12-pdr. Napoleon Brass
	6	1731	7-in. Army
		1732	12-pdr. Napoleon Brass
		1733	4.62-in. Rifle
		1734	10-pdr. Parrott
	13	1735	10-in. Columbiad
	12	1736	12-pdr. Mountain How.
		1737	12-pdr. Mountain How.
		1738	12-pdr. Napoleon Bronze
	15	1739	10-in. Columbiad
		1740	7-in. Brooke Navy
		1741	12-pdr. Napoleon
		1742	12-pdr. Mountain How.
		1743	12-pdr. Mountain How.
	20	1744	12-pdr. Napoleon Brass
	22	1745	10-in. Mortar
		1746	20-pdr. Parrott
		1747	12-pdr. Napoleon Brass
		1748	10-in. Columbiad
		1749	10-in. Mortar
		1750	20-pdr. Parrott
		1751	7-in. Brooke
	27	1752	20-pdr. Parrott
		1753	20-pdr. Parrott
		1754	12-pdr. Napoleon
		1755	12-pdr. Napoleon
	29	1756	10-in. Mortar
	31	1757	7-in. Rifle Army
Feb.	4	1758	7-in. Rifle Army
	6	1759	20-pdr. Parrott
		1760	20-pdr. Parrott
		1761	20-pdr. Parrott
		1762	8-in. Mortar Light
	8	1763	10-in. Columbiad
	14	1764	6.40-in. Brooke
	16	1765	6.40-in. Brooke
		1766	10-in. Columbiad
		1767	10-in. Columbiad
		1768	12-pdr. Napoleon
		1769	12-pdr. Napoleon
		1770	12-pdr. Napoleon
	26	1771	10-in. Columbiad
		1772	10-in. Columbiad
		1773	7-in. Brooke Double Banded
		1774	12-pdr. Napoleon
		1775	12-pdr. Napoleon
		1776	10-in. Columbiad
		1777	12-pdr. Napoleon
		1778	12-pdr. Napoleon
Mar.	3	1779	6.40-in. Brooke
	4	1780	10-in. Columbiad
	7	1781	10-in. Columbiad
		1782	12-pdr. Napoleon
		1783	12-pdr. Napoleon

		1784	12-pdr. Napoleon
		1785	12-pdr. Napoleon
		1786	4.62-in. Napoleon
		1787	4.62-in. Napoleon
	16	1788	10-in. Columbiad
		1789	7-in. Brooke Double Banded
		1790	12-pdr. Napoleon
		1791	12-pdr. Napoleon
		1792	12-pdr. Napoleon
		1793	12-pdr. Napoleon
		1794	10-in. Columbiad
		1795	12-pdr. Napoleon
		1796	12-pdr. Napoleon
		1797	12-pdr. Napoleon
		1798	12-pdr. Napoleon
		1799	12-pdr. Napoleon
		1800	12-pdr. Napoleon
		1801	12-pdr. Napoleon
		1802	12-pdr. Napoleon
		1803	12-pdr. Napoleon
		1804	12-pdr. Napoleon
		1805	12-pdr. Napoleon
		1806	12-pdr. Napoleon
		1807	12-pdr. Napoleon
		1808	12-pdr. Napoleon
		1809	12-pdr. Napoleon
		1810	8-in. Mortar Light
		1811	10-pdr. Parrott
		1812	12-pdr. Mountain How.
		1813	12-pdr. Napoleon
		1814	12-pdr. Napoleon
		1815	12-pdr. Napoleon
		1816	12-pdr. Napoleon
		1817	12-pdr. Napoleon
		1818	12-pdr. Napoleon
		1819	10-pdr. Parrott
Apr.	18	1820	10-in. Mortar
		1821	10-in. Mortar
		1822	12-pdr. Napoleon
		1823	12-pdr. Napoleon
		1824	12-pdr. Napoleon
		1825	12-pdr. Napoleon
		1826	10-in. Mortar
		1827	12-pdr. Napoleon
		1828	12-pdr. Napoleon
		1829	10-in. Mortar
May	1	1830	10-in. Columbiad
		1831	12-pdr. Napoleon
		1832	12-pdr. Napoleon
		1833	12-pdr. Napoleon
		1834	12-pdr. Napoleon
		1835	12-pdr. Napoleon
		1836	12-pdr. Napoleon
		1837	10-in. Mortar
	28	1838	20-pdr. Parrott
		1839	12-pdr. Napoleon
June	3	1841	6.40-in.
		1842	20-pdr. Parrott
	9	1843	6.40-in. Brooke
		1844	12-pdr. Napoleon
		1845	12-pdr. Napoleon
		1846	12-pdr. Napoleon
		1847	12-pdr. Napoleon
		1848	12-pdr. Napoleon
		1849	12-pdr. Napoleon
		1850	12-pdr. Napoleon
	17	1851	10-in. Columbiad
		1852	10-in. Columbiad
		1853	6.40-in. Brooke
		1854	6.40-in. Brooke
		1855	12-pdr. Napoleon
		1856	12-pdr. Napoleon
		1857	12-pdr. Napoleon
		1858	12-pdr. Napoleon
		1859	20-pdr. Parrott
		1860	12-pdr. Napoleon
		1861	12-pdr. Napoleon
		1862	12-pdr. Napoleon
		1863	12-pdr. Napoleon
		1864	12-pdr. Napoleon
July	17	1865	10-in. Columbiad
		1866	10-in. Columbiad
	25	1867	7-in. Brooke Double Banded
		1868	30-pdr. Parrott
		1869	12-pdr. Napoleon
		1870	12-pdr. Napoleon
		1871	12-pdr. Napoleon
		1872	10-in. Columbiad
	30	1873	10-in. Columbiad
		1874	10-in. Columbiad
		1875	12-pdr. Napoleon
		1876	12-pdr. Napoleon
		1877	12-pdr. Napoleon
		1878	30-pdr. Parrott
	27	1879	7-in. Brooke Double Banded
Aug.	8	1880	7-in. Brooke Double Banded
		1881	30-pdr. Parrott
		1882	12-pdr. Napoleon
		1883	12-pdr. Napoleon
		1884	12-pdr. Napoleon
		1885	10-in. Columbiad
	24	1886	7-in. Brooke
		1887	30-pdr. Parrott
		1888	30-pdr. Parrott
		1889	10-in Columbiad
		1890	12-pdr. Napoleon
		1891	12-pdr. Napoleon
		1892	12-pdr. Napoleon
		1893	12-pdr. Napoleon
		1894	12-pdr. Napoleon

		1895	12-pdr. Napoleon
		1896	10-in. Columbiad
Sept.	7	1897	7-in. Brooke
		1898	10-pdr. Parrott
		1899	10-pdr. Parrott
		1900	20-pdr. Parrott
		1901	20-pdr. Parrott
		1902	8-in. Columbiad
		1903	8-in. Columbiad
		1904	10-in. Columbiad
		1905	10-in. Columbiad
		1906	8-in. Columbiad
		1907	12-pdr. Napoleon Brass
		1908	12-pdr. Napoleon Brass
		1909	7-in. Army
		1910	8-in. Columbiad
		1911	10-in. Columbiad
		1912	10-in. Brooke Double Banded
	24	1913	8-in. Columbiad
		1914	12-pdr. Napoleon
		1915	12-pdr. Napoleon
		1916	12-pdr. Napoleon
		1917	12-pdr. Napoleon
		1918	10-in. Columbiad
		1919	10-in. Columbiad
Oct.	1	1920	8-in. Columbiad
		1921	10-in. Columbiad
		1922	8-in. Columbiad
		1923	10-in. Navy Double Banded
	19	1924	10-in. Columbiad
		1925	10-in. Columbiad
		1926	10-in. Columbiad
		1927	12-pdr. Napoleon
		1928	12-pdr. Napoleon
		1929	12-pdr. Napoleon
		1930	12-pdr. Napoleon
	13	1931	7-in. Navy Double Banded
		1932	8-in. Columbiad
		1933	30-pdr. Parrott
		1934	12-pdr. Napoleon
		1935	12-pdr. Napoleon
		1936	12-pdr. Napoleon
		1937	12-pdr. Napoleon
		1938	10-in. Navy Double Banded
		1939	30-pdr. Parrott
	18	1940	20-pdr. Parrott
		1941	20-pdr. Parrott
		1942	10-pdr. Parrott
		1943	10-pdr. Parrott
		1944	10-in. Columbiad
		1945	8-in. Columbiad
	27	1946	7-in. Army
		1947	8-in. Columbiad
	24	1948	20-pdr. Parrott
		1949	20-pdr. Parrott
		1950	10-pdr. Parrott
		1951	10-pdr. Parrott
	29	1952	10-in. Columbiad
		1953	8-in. Siege
		1954	12-pdr. Napoleon
		1955	12-pdr. Napoleon
		1956	12-pdr. Napoleon
Nov.	2	1957	10-in. Navy Double Banded
		1958	8-in. Columbiad
	5	1959	10-in. Columbiad
		1960	30-pdr. Parrott
	6	1961	20-pdr. Parrott
		1962	20-pdr. Parrott
		1963	10-pdr. Parrott
		1964	10-pdr. Parrott
	9	1965	10-in. Columbiad
		1966	8-in. Siege
		1967	20-pdr. Parrott
		1968	20-pdr. Parrott
		1969	10-pdr. Parrott
		1970	10-pdr. Parrott
		1971	10-pdr. Parrott
		1972	10-pdr. Parrott
		1973	20-pdr. Parrott
		1974	20-pdr. Parrott
	13	1975	10-in. Columbiad
		1976	7-in. Navy Brooke
		1977	10-in. Columbiad
	16	1978	30-pdr. Parrott
		1979	10-in. Columbiad
		1980	8-in. Siege
		1981	10-in. Columbiad
		1982	7-in. Navy
		1983	20-pdr. Parrott
		1984	20-pdr. Parrott
		1985	20-pdr. Parrott
		1986	10-pdr. Parrott
		1987	10-pdr. Parrott
		1988	10-pdr. Parrott
		1989	10-pdr. Parrott
		1990	10-pdr. Parrott
		1991	10-pdr. Parrott
		1992	10-pdr. Parrott
		1993	10-pdr. Parrott
		1994	12-pdr. Napoleon
		1995	12-pdr. Napoleon
		1996	12-pdr. Napoleon
		1997	12-pdr. Napoleon
		1998	12-pdr. Napoleon
		1999	10-in. Columbiad
		2000	8-in. Columbiad
		2001	10-in. Columbiad
		2002	30-pdr. Parrott
Dec.	7	2003	10-in. Columbiad
		2004	7-in. Navy

	10	2005	10-in. Columbiad
		2006	8-in. Columbiad
		2007	8-in. Columbiad
		2008	10-in. Columbiad
		2009	8-in. Columbiad
		2010	7-in.
		2011	11-in. Brooke
		2012	8-in. Columbiad
		2013	10-in. Columbiad
		2014	10-pdr. Parrott
		2015	10-pdr. Parrott
		2016	10-pdr. Parrott
		2017	10-pdr. Parrott
			1864
Jan.	14	2018	12-pdr. Napoleon Iron Defective
		2019	12-pdr. Napoleon Iron Defective
		2020	12-pdr. Napoleon Iron Defective
		2021	12-pdr. Napoleon Iron Defective
	18	2022	12-pdr. Napoleon Iron Defective
		2023	12-pdr. Napoleon Iron Defective
		2024	12-pdr. Napoleon Iron Heads Cut
		2025	12-pdr. Napoleon Iron 2½ in. short
	22	2026	12-pdr. Napoleon Iron
		2027	12-pdr. Napoleon Iron
		2028	12-pdr. Napoleon Iron
		2029	12-pdr. Napoleon Iron
	27	2030	12-pdr. Napoleon Iron
		2031	12-pdr. Napoleon Iron
		2032	12-pdr. Napoleon Iron
		2033	12-pdr. Napoleon Iron
Feb.	3	2034	7-in. Brooke
		2035	8-in. Siege
	2	2036	12-pdr. Napoleon Iron
		2037	12-pdr. Napoleon Iron
		2038	12-pdr. Napoleon Iron Rough
		2039	12-pdr. Napoleon Iron Rough
	9	2040	11-in. Brooke
		2041	12-pdr. Napoleon Iron
		2042	12-pdr. Napoleon Iron
		2043	12-pdr. Napoleon Iron
		2044	12-pdr. Napoleon Iron
	11	2045	10-in. Columbiad
		2046	8-in. Siege
		2047	12-pdr. Napoleon Iron
		2048	12-pdr. Napoleon Iron
		2049	12-pdr. Napoleon Iron
		2050	12-pdr. Napoleon Iron
		2051	7-in. Brooke
		2052	8-in. Siege
	19	2053	12-pdr. Napoleon Iron
		2054	12-pdr. Napoleon Iron
		2055	12-pdr. Napoleon Iron
		2056	12-pdr. Napoleon Iron
	21	2057	12-pdr. Napoleon Iron
		2058	12-pdr. Napoleon Iron Rejected
		2059	12-pdr. Napoleon Iron
		2060	12-pdr. Napoleon Iron
	19	2061	10-in. Columbiad
		2062	6.40-in. Brooke
		2063	7-in. Brooke
	24	2064	6.40-in. Brooke
		2065	10-in. Columbiad
Mar.	11	2066	10-in. Columbiad
	28	2067	7-in. Navy
		2068	10-in. Mortar
	21	2069	10-in. Mortar
		2070	8-in. Siege
	28	2071	6.40-in. Brooke
Apr.	1	2072	10-pdr. Parrott
		2073	10-pdr. Parrott
		2074	10-pdr. Parrott
		2075	10-pdr. Parrott
	6	2076	8-in. Rifle Navy
		2077	8-in. Siege
	7	2078	8-in. Siege
		2079	10-in. Columbiad
	8	2080	12-pdr. Napoleon Iron
		2081	12-pdr. Napoleon Iron
	9	2082	10-in. Columbiad
	11	2083	12-pdr. Napoleon Iron
		2084	12-pdr. Napoleon Iron
		2085	12-pdr. Napoleon Iron
		2086	12-pdr. Napoleon Iron
		2087	7-in. Brooke
	13	2088	20-pdr. Parrott
		2089	12-pdr. Napoleon
		2090	12-pdr. Napoleon
	14	2091	10-pdr. Parrott
		2092	10-pdr. Parrott
		2093	20-pdr. Parrott
		2094	20-pdr. Parrott
	16	2095	8-in. Brooke Rifle
		2096	8-in. Siege
		2097	12-pdr. Napoleon Iron
		2098	12-pdr. Napoleon Iron
		2099	10-pdr. Parrott
		2100	10-pdr. Parrott
	19	2101	20-pdr. Parrott
		2102	20-pdr. Parrott
		2103	10-pdr. Parrott
	20	2104	10-in. Columbiad
	25	2105	7-in. Brooke
		2106	10-in. Columbiad
		2107	8-in. Rifle
	28	2108	8-in. Siege
May	2	2109	10-in. Columbiad
		2110	8-in. Siege
	27	2111	7-in. Brooke Navy

		2112	6.40-in. Brooke Navy
June	1	2113	10-in. Columbiad
		2114	8-in. Siege
		2115	8-in. Rifle Brooke
		2116	6.40-in. Brooke
	8	2117	12-pdr. Napoleon Iron
		2118	12-pdr. Napoleon Iron
		2119	10-pdr. Parrott
		2120	10-pdr. Parrott
	11	2121	10-in. Columbiad
		2122	8-in. Siege
	13	2123	12-pdr. Napoleon Iron
		2124	12-pdr. Napoleon Iron
		2125	12-pdr. Napoleon Iron
		2126	12-pdr. Napoleon Iron
	16	2127	10-in. Columbiad
		2128	8-in. Siege
	18	2129	10-pdr. Parrott
		2130	10-pdr. Parrott
		2131	12-pdr. Napoleon Iron
		2132	12-pdr. Napoleon Iron
		2133	10-in. Columbiad
	21	2134	6.40-in. Brooke
	23	2135	12-pdr. Napoleon
		2136	12-pdr. Napoleon
		2137	12-pdr. Napoleon
		2138	12-pdr. Napoleon
	27	2139	7-in. Brooke
		2140	8-in. Siege
	29	2141	12-pdr. Napoleon
		2142	12-pdr. Napoleon
		2143	12-pdr. Napoleon
		2144	12-pdr. Napoleon
		2145	12-pdr. Napoleon
July	1	2146	10-in. Columbiad
		2147	7-in. Brooke Not rifled or banded
		2148	8-in. Siege
	4	2149	12-pdr. Napoleon
		2150	12-pdr. Napoleon
		2151	12-pdr. Napoleon
		2152	6.40-in. Brooke
	12	2153	30-pdr. Parrott
		2154	30-pdr. Parrott
		2155	8-in. Mortar
		2156	8-in. Mortar
		2157	30-pdr. Parrott
		2158	30-pdr. Parrott
Aug.	25	2159	10-pdr. Parrott
		2160	10-pdr. Parrott
		2161	12-pdr. Napoleon
		2162	12-pdr. Napoleon
	30	2163	12-pdr. Napoleon
		2164	12-pdr. Napoleon
		2165	10-pdr. Parrott
		2166	10-pdr. Parrott
Sept.	2	2167	12-pdr. Napoleon
		2168	12-pdr. Napoleon
		2169	10-pdr. Parrott
		2170	10-pdr. Parrott
	6	2171	12-pdr. Napoleon
		2172	12-pdr. Napoleon
		2173	12-pdr. Napoleon
		2174	12-pdr. Napoleon
	10	2175	12-pdr. Napoleon
		2176	12-pdr. Napoleon
		2177	12-pdr. Napoleon
		2178	12-pdr. Napoleon
	13	2179	10-in. Seacoast Mortar
	15	2180	10-pdr. Parrott
		2181	12-pdr. Napoleon
	23	2182	12-pdr. Napoleon
		2183	10-Napoleon [?]
	17	2184	10-in. Seacoast Mortar
Oct.	22	2185	12-pdr. Napoleon
		2186	12-pdr. Napoleon
		2187	8-in. Mortar Siege
	25	2188	8-in. Mortar Siege
		2189	12-pdr. Napoleon
		2190	12-pdr. Napoleon
	29	2191	12-pdr. Napoleon
		2192	12-pdr. Napoleon
		2193	8-in. Mortar
Nov.	1	2194	10-pdr. Parrott
		2195	12-pdr. Napoleon
		2196	12-pdr. Napoleon
		2197	8-in. Mortar
	4	2198	8-in. Mortar
		2199	12-pdr. Napoleon
		2200	12-pdr. Napoleon
	10	2201	8-in. Mortar
		2203	12-pdr. Napoleon
		2204	12-pdr. Napoleon
	17	2205	8-in. Mortar
		2206	12-pdr. Napoleon
		2207	12-pdr. Napoleon
	14	2208	12-in. Columbiad Cast Hollow
		2209	10-in. Mortar
	22	2210	20-pdr. Parrott
		2211	10-pdr. Parrott
		2212	10-pdr. Parrott
	29	2213	10-in. Columbiad
	30	2214	20-pdr. Parrott
		2215	12-pdr. Napoleon
		2216	12-pdr. Napoleon
Dec.	6	2217	20-pdr. Parrott
		2218	12-pdr. Napoleon
		2219	12-pdr. Napoleon
	2	2220	10-in. Columbiad
	12	2221	20-pdr. Parrott
		2223	12-pdr. Napoleon

		2224	12-pdr. Napoleon
		2225	8-in. Mortar
		2226	8-in. Seacoast Mortar
		2227	10-in. Columbiad
		2228	8-in. Seacoast Mortar
	17	2229	8-in. Siege Mortar
		2230	12-pdr. Napoleon
		2231	12-pdr. Napoleon
		2232	10-in. Columbiad
			1865
Jan.	12	2234	12-pdr. Napoleon
		2235	12-pdr. Napoleon
		2236	24-pdr.
		2237	24-pdr.
	16	2238	7-in. Brooke
		2239	10-in. Columbiad
	19	2240	24-pdr.
		2241	24-pdr.
		2242	12-pdr. Napoleon
		2243	12-pdr. Napoleon
		2244	12-pdr. Napoleon
		2245	12-pdr. Napoleon
		2246	12-pdr. Napoleon
		2247	12-pdr. Napoleon
		2248	24-pdr.
		2249	12-pdr. Napoleon
		2250	12-pdr. Napoleon
		2251	12-pdr. Napoleon
Feb.	20	2252	12-in. Columbiad Cast Hollow
		2253	24-pdr.
		2254	10-pdr. Parrott
		2255	10-pdr. Parrott
		2256	12-pdr. Napoleon
		2257	10-pdr. Parrott
		2258	12-pdr. Napoleon
		2250	12-pdr. Napoleon
		2260	7-in. Brooke
		2261	10-in. Mortar
		2262	24-pdr.
		2263	12-pdr. Napoleon
		2264	12-pdr. Napoleon
March		2265	12-pdr. Napoleon
		2266	12-pdr. Napoleon
		2267	12-pdr. Napoleon
		2268	12-pdr. Napoleon

CANNON CHANNELED THROUGH THE ATLANTA ARSENAL

Between August, 1863, and June, 1864, some 201 cannon were channeled through the arsenal in Atlanta, Georgia. Arsenal clerks recorded some or all cannon inscriptions for 142 of these weapons (two of which are illegibly written). Clerks either failed to record the inscriptions for the remaining cannon or the weapons were without markings. Doubtless many of the latter group were Confederate-made. The following is a Table of those weapons for which some markings were transcribed. The Table includes sixty-five cannon with definite Confederate foundry markings. Calibers and markings are, for the most part, extracted exactly as transcribed in the original. Occasional interpretations are made for obvious errors, however.

The weapons listed as having passed through the Arsenal on October 30, 1863, were captured by the Confederates at the Battle of Chickamauga in September. The weapons that are listed as having passed through on January 26 and 28, April 11 and 19, and June 29 and 30 were all fowarded to the Augusta Arsenal. The probability is that most of the guns were sent there for scrapping.

CANNON CHANNELED THROUGH THE ATLANTA ARSENAL
August 1863-July 1864

DATE	TYPE	MAKER
26 Aug. 1863	6-pdr.	Quinby & Robinson Memphis, Tenn.
	6-pdr.	A.B.R. & Bro. Vicksburg, Miss. No. 16
	6-pdr. Iron	T. M. Brennan Nashville, Tenn. No. 5 887 lbs. MHW

DATE	TYPE	MAKER
16 Oct.	12-pdr. How.	J.R.A. & Co. No. 1207
	12-pdr. How.	C.A. & Co.
30 Oct.	3.8-in. Bronze Rifle	N. P. Ames Springfield, Mass. No. 97
	3.8-in. Bronze Rifle	Greenwood & Co. Cin., O. No. 54
	12-pdr. Mt. How.	C.A. & Co. 220 lbs. No. 204 1862
	12-pdr. Mt. How.	J.R.A. & Co. No. 1555 222 lbs. 1862
	12-pdr. Mt. How.	C.A. & Co. Boston No. 29 1846
	12-pdr. Mt. How.	C.A. & Co. Boston No. 63 1860
	12-pdr. Iron How.	J.R.A. & Co. 1861
	12-pdr. Iron How.	J.R.A. & Co. 1862
	3-in. Iron Rifle	J.R.A. & Co. No. 1480 908 lbs.
	6-pdr.	C.A. & Co. Boston No. 195
	12-pdr. How.	N. P. Ames Springfield, Mass. No. 86
	12-pdr. Iron How.	J.R.A. & Co. 1861
	24-pdr. How.	Maker: ? No. 11 660 lbs. 1857
	24-pdr. How.	Maker: ? No. 44 674 lbs. 1859
	3.8-in. Bronze Rifle	M. Greenwood Cin., O. No. 37
	3.8-in. Bronze Rifle	Ames & Co. Chicopee, Mass. 1861 864 lbs.
	3.8-in. Bronze Rifle	Ames & Co. Chicopee, Mass. 1861 873 lbs.
	12-pdr. How.	Ames & Co. Chicopee, Mass. 1856 760 lbs.
	3.8-in. Bronze Rifle	C.A. & Co. 879 lbs.
	12-pdr. How.	M. Greenwood Cin., O. No. 56 1861
	3.8-in. Bronze Rifle	M. Greenwood Cin., O. No. 28 1862 854 lbs.
	3.8-in. Bronze Rifle	N. P. Ames Springfield, Mass. No. 69
	12-pdr. Napoleon	Revere Copper & Co. 1862 1242 lbs. No. 142
	12-pdr. How.	Alleghany Arsenal No. 7 1858
	6-pdr.	C.A. & Co. Boston
	6-pdr.	M. Greenwood & Co. Cin., O. 1861
	12-pdr. How.	Marshall & Co. St. Louis 1862
	6-pdr.	M. Greenwood Cin., O. 1861
	6-pdr.	M. Greenwood Cin., O. 1861
	12-pdr. Napoleon	M. Greenwood & Co., Cin., O. 1224 lbs. 1862
	3-in. Steel Rifle	P.I. Co. 817 lbs. No. 86

DATE	TYPE	MAKER
	3.8-in. Bronze Rifle	A.M. Co. 1862 No. 56 924 lbs.
	6-pdr.	C.A. & Co. Boston
	12-pdr. Mt. How.	C.A. & Co. Boston No. 201 1862
	12-pdr. How.	John Clark & Co. N.O.
	12-pdr. How.	C.A. & Co. 1862 No. 186
	12-pdr. How.	Marshall & Co. St. Louis No. 16
	6-pdr.	N. P. Ames Springfield, Mass. No. 38
	6-pdr.	C.A. & Co. Boston No. 126 1853
	12-pdr. How.	Noble & Bro. Rome, Ga. 1862
	12-pdr. Mt. How.	C.A. & Co. No. 197
	12-pdr. Mt. How.	J.R.A. & Co. 1862
	3-in. Rifle	Noble & Bro. Rome, Ga.
	3-in. Rifle	Noble & Bro. Rome, Ga. 1862
	12-pdr. How.	M. Greenwood & Co., Cin., O. No. 92
	12-pdr. How.	A.B.R. & Bro. Vicksburg, Miss
5 Nov.	12-pdr. Napoleon	M. Greenwood No. 40 1268 lbs.
	12-pdr. Napoleon	M. Greenwood No. 19 1224 lbs.
	12-pdr. Napoleon	Revere Copper Co. No. ? 1247 lbs.
11 Nov.	12-pdr. Hows. (4)	J.R.A. & Co.
19 Nov.	12-pdr. Napoleon	Ga. R.R. Machine Shop Augusta No. 37 1207 lbs. GWR
26 Jan. 1864	12-pdr. How.	Noble Bros. Rome, Ga.
28 Jan.	Breechloading Gun	Rushton Georgia Rail Road Atlanta, Ga.
3 Feb.	12-pdr. How.	Ames & Co. Chicopee, Mass. No. 31 768 lbs.
	12-pdr. How.	C.A. & Co. Boston No. 15 768 lbs.
18 Feb.	12-pdr. How.	N. P. Ames Springfield, Mass. No. 49 785 lbs.
	12-pdr. How.	C.A. & Co. Boston 1839
1 March	12-pdr. Napoleon	Columbus Arsenal No. 25 1216 lbs.
	12-pdr. Napoleon	Columbus Arsenal No. 26 1212 lbs.
	12-pdr. Napoleon	Columbus Arsenal No. 27 1228 lbs.
	12-pdr. Napoleon	Columbus Arsenal No. 28 1227 lbs.
14 March	12-pdr. How.	C.A. & Co. Boston
	12-pdr. How.	C.A. & Co. Boston No. 179 798 lbs.
	12-pdr. How.	Ames Co. Chicopee, Mass. No. 139

DATE	TYPE	MAKER
	12-pdr. How.	J.R.A. & Co. No. 1207
24 March	12-pdr. How.	N. P. Ames Springfield, Mass. No. 10 787 lbs.
	12-pdr. How.	J.R.A. & Co. No. 1529 765 lbs.
	12-pdr. How.	W.J.H. W.F. S.&P. STATE VA.
	12-pdr. How.	M. Greenwood Cin., O. No. 78 786 lbs.
31 March	12-pdr. Napoleon	Augusta Foundry No. 13 GWR
6 April	12-pdr. Napoleon	Macon Arsenal No. 41 1164 lbs.
	12-pdr. Napoleon	Macon Arsenal No. 42 1152 lbs.
	12-pdr. Napoleon	Macon Arsenal No. 43 1184 lbs.
	12-pdr. Napoleon	Macon Arsenal No. 44 1175 lbs.
7 April	12-pdr. Mt. How.	C.A. & Co. No. 198 219 lbs.
	12-pdr. Mt. How.	C.A. & Co. No. 199 223 lbs.
	12-pdr. Mt. How.	C.A. & Co. No. 224 222 lbs.
	12-pdr. Mt. How.	C.A. & Co. No. 222 220 lbs.
11 April	3-in. Bronze Breechloading	Georgia Rail Road Machine Shop
	6-pdr.	Ames & Co. Chicopee, Mass.
	6-pdr. Rifle	Greenwood & Co. Cin., O.
	6-pdr.	C.A. & Co. Boston, Mass.
	6-pdr. Iron	Noble Bros. & Co. Rome, Ga.
19 April	6-pdr.	A.B.R. & Bro. Vicksburg No. 14 1862
	6-pdr.	C.A. & Co. Boston No. 1
	6-pdr.	C.A. & Co. Boston No. 34
	6-pdr.	Greenwood 855 lbs.
	6-pdr.	Ames & Co.
	3.26-in. Bronze Rifle	Leeds & Co. 893 lbs. 1861
	3.29-in. Bronze Rifle	Leeds & Co. 887 lbs. 1861
	3.75-in. Bronze Rifle	Ames & Co. 870 lbs.
20 April	12-pdr. Mt. How.	C.A. & Co. No. 202 225 lbs.
	12-pdr. Mt. How.	C.A. & Co. No. 220 221 lbs.
25 April	20-pdr. Parrott	J.R.A. & Co. No. 1967 1864 1858 lbs.
	20-pdr. Parrott	J.R.A. & Co. No. 1968 1864 1869 lbs.
3 May	12-pdr. Napoleon	Augusta Foundry No. 77 1220 lbs.
	3-in. Iron Rifle	Noble Bros. Rome, Ga.
4 May	12-pdr. Napoleon	Augusta Foundry No. 73 1205 lbs.
	12-pdr. Napoleon	Augusta Foundry No. 74 1207 lbs.
	12-pdr. Napoleon	Augusta Foundry No. 75 1206 lbs.

DATE	TYPE	MAKER
	12-pdr. Napoleon	Augusta Foundry No. 76 1212 lbs.
20 May	24-pdr. Siege	C.A. & Co. Boston 5758 lbs.
2 June	10-pdr. Parrott	Macon Arsenal No. 11 868 lbs. EF
3 June	12-pdr. Napoleon	Augusta Foundry 1195 lbs.
	12-pdr. Napoleon	Augusta Foundry 1195 lbs.
	12-pdr. Napoleon	Columbus Arsenal No. 27 1228 lbs. FCH
6 June	6-pdr.	Leeds & Co. N.O.
	3-in. Bronze Rifle	Noble Bros. & Co. Rome, Ga.
	6-pdr. Iron	T. M. Brennan Nashville, Tenn.
15 June	6-pdr.	Ames & Co. Chicopee, Mass.
	6-pdr.	C.A. & Co.
	6-pdr.	Noble Bros. & Co. Rome, Ga. 926 lbs.
	6-pdr.	A.B.R. & Bros. Vicksburg No. 9
29 June	12-pdr. Mt. How.	C.A. & Co. Boston No. 63 219 lbs.
	12-pdr. Mt. How.	J.R.A. & Co. No. 1616 218 lbs.
	12-pdr. Mt. How.	J.R.A. & Co. No. 1585 220 lbs.
	3-in. Iron Rifle	Noble Bros. & Co. Rome, Ga.
	James Rifle	Ames & Co. Chicopee, Mass.
	James Rifle	Ames & Co. Chicopee, Mass.
	3-in. Iron Rifle	Noble Bros. & Co. Rome, Ga. 863 lbs.
	6-pdr. Iron	T.M. Brennan Nashville, Tenn. No. 28
	6-pdr. Iron	T.M. Brennan Nashville, Tenn. No. 59
	6-pdr.	Ames & Co. Chicopee, Mass. 880 lbs.
	6-pdr.	Ames & Co. Chicopee, Mass. No. 381 885 lbs.
	6-pdr.	N. P. Ames Springfield, Mass. No. 128 886 lbs.
	6-pdr.	Quinby & Robinson No. 17
	6-pdr.	C.A. & Co. No. 77 892 lbs.
	6-pdr.	N. P. Ames Springfield, Mass. No. 356 884 lbs.
	Williams Mt. Rifle (2)	Nos. 13 and 15 (Probably made at Tredegar)
	6-pdr.	N. P. Ames Springfield, Mass. No. 43
	6-pdr.	C.A. & Co. No. 110
	12-pdr. Napoleon	Columbus Arsenal No. 7 1205 lbs. FCH CSA
30 June	10-pdr. Parrott	W.P.F. R.P.P. No. 48
	12-pdr. Napoleon	Augusta Foundry No. 3
	12-pdr. Napoleon	Macon Arsenal No. 56

GUNS IN WALKER'S DIVISION, ARMY OF MISSISSIPPI
1863 (Month Unknown)

TYPE	MAKER
12-pdr. Napoleon	Macon Arsenal No. 5 1210 lbs.
12-pdr. Napoleon	Macon Arsenal No. 6 1201 lbs.
12-pdr. Napoleon	Macon Arsenal No. 9 1188 lbs.
12-pdr. Napoleon	Macon Arsenal No. 10 1185 lbs.
12-pdr. How.	without marks
12-pdr. How.	without marks
12-pdr. How.	Columbus, Ga. 788 lbs.
12-pdr. How.	Columbus, Ga. 788 lbs.
12-pdr. How.	Noble Bros. & Co. Rome, Ga.
12-pdr. How.	Noble Bros. & Co. Rome, Ga.
12-pdr. How.	J.R.A. & Co. No. 1447 775 lbs.
12-pdr. How. Iron	J.R.A. & Co. No. 1449
12-pdr. How.	C.A. & Co. Boston, Mass. No. 10
12-pdr. How.	
6-pdr.	N. P. Ames No. 322 884 lbs.
6-pdr.	
6-pdr.	John Clark New Orleans
6-pdr.	N. P. Ames Springfield, Mass. No. 4 802 lbs.
*20-pdr. Parrott	W.P. Foundry R.P.P. No. 35 1695 lbs.
20-pdr. Parrott	J. R. A. & Co. No. 1726
20-pdr. Parrott	J.R.A. & Co. No. 1712
3-in. Parrott	
3-in. Parrott	
3-in. Bronze Rifle	A.B.R. & Bro. Vicksburg, Miss. No. 32 775 lbs.

*Captured at Corinth, Miss., on Oct. 5, 1862

CANNON ABANDONED BY THE CONFEDERATES IN THE PETERSBURG TRENCHES

TYPE	MAKER
3-in. Parrott	J.R.A. & Co. No. 2180
3-in. Parrott	J.R.A. & Co. No. 2170
3-in. Parrott	W.P.F. R.P.P. No. 95
3-in. Parrott	U.S. no marks
4½-in Iron How.	J.R.A. & Co.
24-pdr. Boat How.	Ames Manufacturing Company No. 111 1297-98
12-pdr. Napoleon	Ames Manufacturing Company No. 33 1862
12-pdr. Napoleon	J.R.A. & Co. No. 1817 1220 lbs.
12-pdr. Napoleon	J.R.A. & Co. No. 1818 1260 lbs.
12-pdr. Napoleon	J.R.A. & Co. No. 1802 1250 lbs.
12-pdr. Napoleon	J.R.A. & Co. No. 1797 1245 lbs.
12-pdr. Napoleon	J.R.A. & Co. No. 1814 1255 lbs.
12-pdr. Napoleon	J.R.A. & Co. No. 1828 1245 lbs. 1863
12-pdr. Napoleon	J.R.A. & Co. No. 1858 1225 lbs. 1863
12-pdr. Napoleon, Iron	J.R.A. & Co. No. 2058
12-pdr. Napoleon, Iron	J.R.A. & Co. No. 2118
12-pdr. Napoleon, Iron	J.R.A. & Co. no number
12-pdr. Napoleon, Iron	J.R.A. & Co. No. 2126
12-pdr. Napoleon, Iron	J.R.A. & Co. No. 2126 [sic]
12-pdr. Napoleon, Iron	J.R.A. & Co. No. 2054

BIBLIOGRAPHY

MANUSCRIPT COLLECTIONS

Account Book, 1861–63, of Edward R. Archer, Virginia Historical Society.

Daily Report of Ordnance and Ordnance Stores on Hand, Macon Arsenal, 1863–64, Chap. IV, Vol. 58, National Archives Record Group 109 (War Department's Collection of Confederate Records).

House Files of the Southern Claims Commission, National Archives.

Letters Received, Macon Arsenal, Chap. IV, Vol. 4, National Archives Record Group 109 (War Department's Collection of Confederate Records).

Letters Received, Secretary of War, National Archives Record Group 109 (War Department's Collection of Confederate Records).

Letters Relating to A. N. Miller and Noble Brothers & Company, Ordnance Department of Georgia, Georgia State Archives.

Listing of Cannon in Walker's Division, Army of Mississippi, 1863 (month unknown), Mississippi State Archives.

Military Board Record Book, Vol. 63, Series No. 6, Record Group 21, Tennessee State Archives.

Papers Relating to Citizens or Business Firms, National Archives Record Group 109 (War Department's Collection of Confederate Records).

Phillip Crutcher Collection, Mississippi State Archives.

Records of Receipts and Deliveries, Ordnance Office, Atlanta, Georgia, 1863, Chap. IV, Vol. 88, National Archives Record Group 109 (War Department's Collection of Confederate Records).

Record of Receipts and Issues at the Atlanta Armory, Dec. 1863–June 1864, Vols. 17–18, National Archives Record Group 109 (War Department's Collection of Confederate Records).

Record of Operations of the Selma Naval Ordnance Works, 2 Vols., National Archives Record Group 45 (Naval Records).

Tredegar Foundry Sales Book 1860–67, Virginia State Library.

Tredegar Gun Foundry Book 1861–65, Virginia State Library.

Union Provost Marshal's "Citizens File" pertaining to Samuel Noble, National Archives.

OFFICIAL PUBLICATIONS

Chandler, Allen D. *The Confederate Records of the State of Georgia,* 2 Vols. Atlanta: Charles P. Byrd, 1909.

Freeman, Douglas S. *A Calender of Confederate Papers.* Richmond: The Confederate Museum, 1908.

Gardner's New Orleans Directory for 1861.

Gardner's New Orleans Directory for 1866.

Nashville City Directory for 1861.

Official Records of the Union and Confederate Navies in the War of the Rebellion, 31 Vols. Washington: U.S. Government Printing Office, 1894–1927.

U.S. House of Representatives Document No. 99, 2nd Session, 40th Congress, 1867–68. Washington: U.S. Government Printing Office, 1868.

War of the Rebellion: A Compilation of the Official Records of the Union and Confederate Armies, 128 Vols. Washington: U.S. Government Printing Office, 1880–1901.

Williams' Memphis Directory, City Guide, and Business Mirror, 1860.

NEWSPAPERS

Atlanta *Southern Confederacy*
Augusta *Constitutionalist*
Baton Rouge *Daily Advocate*
Charleston *Courier*
Charleston *Mercury*
Columbus *Enquirer*
Columbus *Sun*
Macon *Telegraph*
Memphis *Appeal*
Memphis *Avalanche*
Nashville *Patriot*
Nashville *Union and American*
Natchez *Courier*
New York *Times*
New Orleans *Bee*
New Orleans *Picayune*
New Orleans *Times*
Petersburg *Express*
Philadelphia *Inquirer*
Raleigh *Weekly Register*
Richmond *Dispatch*
Richmond *Examiner*
Rome *Courier Weekly*
Vicksburg *Daily Whig*
Wilmington *Journal*

BOOKS AND ARTICLES

Albaugh, William A. and Edward N. Simmons. *Confederate Arms.* Harrisburg, Pa.: The Stackpole Company, 1957.

———, Hugh Benet, and Edward N. Simmons. *Confederate Handguns.* York, Pa.: George Shumway, 1963.

Alexander, E. P. "Confederate Artillery Service."

Southern Historical Society Papers, XI (1883), 98–112.
———. *Military Memoirs of a Confederate.* Bloomington: Indiana University Press, 1962.

Armes, Ethel. *The Story of Iron and Coal in Alabama.* Birmingham: Chamber of Commerce, 1910.

Batty, George M., Jr. *A History of Rome and Floyd County.* Atlanta: The Webb and Vary Company, 1922.

Bearss, Edwin C. "Artillery Study Shiloh National Military Park." Project No. 17, an unpublished paper.

———, and Warren E. Grabau. "How Porter's Flotilla Ran the Gantlet Past Vicksburg." *Civil War Times Illustrated,* I (Dec. 1962), 38–48.

Brewer, James H. *The Confederate Negro: Virginia Craftsmen and Military Laborers, 1861-1865.* Durham, N.C.: Duke University Press, 1969.

Brown, William LeRoy. "The Red Artillery: Confederate Ordnance During the War." *Southern Historical Society Papers,* XXVI (1898), 365–376.

Bruce, Kathleen. *Virginia Iron Manufacture in the Slave Era.* New York: The Century Company, 1930.

Clark, Walter. *Histories of the Several Regiments and Batteries From North Carolina in the Great War 1861–65.* Raleigh: State of North Carolina, 1901.

Corley, Florence C. *Confederate City: Augusta, Georgia, 1860-1865.* Columbia: University of South Carolina Press, 1960.

Daves, Graham. "Artillery at the Southern Arsenals." *Southern Historical Society Papers,* XII (1884), 360.

Dew, Charles B. *Ironmaker to the Confederacy: Joseph R. Anderson and the Tredegar Iron Works.* New Haven and London: Yale University Press. 1966.

Donelly, Ralph W. "Confederate Copper." *Civil War History,* I (Dec. 1955), 355–370.

Dubose, Beverly M., III. "The Manufacutre of Confederate Ordnance in Georgia." *The Atlanta Historical Bulletin,* XII (1967), 8–19.

Dufour, Charles L. *The Night the War Was Lost.* Garden City: Doubleday and Company, Inc., 1960.

Fremantle, Arthur James L. *Three Months in the Southern States: April-June 1863.* New York: John Bradburn, 1864.

Gorgas, Josiah. "Notes on the Ordnance Department of the Confederate Government." *Southern Historical Society Papers,* XII (1884), 67–94.

Harewell, Richard B., ed. *Cities and Camps of the Confederate States by Fitzgerald Ross.* Urbana: University *of Illinois Press,* 1958.

Hazlett, James C. "Artillery in the Field". Gettysburg Tour: The Lincoln Civil War Society of Philadelphia.

———. "The Confederate Napoleon Gun." *The Military Collector and Historian,* XVI (Winter 1964), 104–110.

———. "The Federal Napoleon Gun." *The Military Collector and Historian,* XV (Winter 1963), 103–108.

———. "The Napoleon Gun, Its Origin and Introduction into American Service." *The Military Collector and Historian,* XV (Spring 1963), 1–5.

———. "The 3-Inch Ordnance Rifle." *Civil War Times Illustrated,* VII (Dec. 1968), 30–36.

———. "The Parrott Rifle." *Civil War Times Illustrated,* V (Nov. 1966), 27–33.

Horn, Stanley F., ed. *Tennessee's War 1861-1865.* Nashville: Tennessee Civil War Centennial Commission, 1965.

Klein, Maury. *Edward Porter Alexander.* Athens, Ga.: University of Georgia Press, 1971.

Lee, Susan P., ed. *The Memoirs of William Nelson Pendleton.* Philadelphia: J. P. Lipponcott Company, 1893.

Lockett, S. H. "The Defense of Vicksburg." *Battles and Leaders of the Civil War,* III, 482–492. New York and London: Thomas Yoseloff, 1956.

McDonald, Ed, Wayne Praeder, Mike Bianchi Rossi. "Rifle Musket: American Firepower and Tactics 1861-1865." *Conflict,* VII (June 1974), 4–26.

McIntosh, David G. "The Confederate Artillery—Its Organization and Development." *The Photographic History of the Civil War,* V, 56–70. New York and London: Thomas Yoseloff, 1957.

———. "Major James H. Warner." Unpublished and unsigned paper in Alabama State Archives.

Mallet, J. W. "Work of the Ordnance Bureau." *Southern Historical Society Papers,* XXXVII (1909), 1–20.

Melton, Maurice. "The Selma Naval Ordnance Works." *Civil War Times Illustrated,* XIV (Dec. 1975), 18–25, 28–31.

Mitchell, Stephen. "Atlanta: The Industrial Heart of the Confederacy." *The Atlanta Historical Bulletin,* (May 1930), 20–27.

Naiswald, L. Van Loan. *Grape and Canister.* New York: Oxford University Press, 1960.

Olmstead, Edwin H. "Bellona Digest and Index." An unpublished paper.

———. "Brooke Digest." An unpublished paper.

Peterson, Harold L. *Round Shot and Rammers.* New York: Bonanza Books, 1969.

Rains, G. W. *History of the Confederate Powder Works.* Augusta: Chronicle and Constitutionalist Press, 1882.

Ripley, Warren. *Artillery and Ammunition of the Civil War.* New York: Van Nostrand Reinhold Company, 1970.

Roberts, Kate Quintard Noble. "A War Time Foundry." *The Alabama Historical Quarterly,* XVIII (1956), 463–473.

Roman, Alfred. *The Military Operations of General Beauregard,* 2 Vols. New York: Harper and Brothers, 1884.

Standard, Diffee W. *Columbus, Georgia, in the Confederacy.* New York: William-Frederick Press, 1954.

Stephens, Walter W. "The Brooke Guns From Selma." *The Alabama Historical Quarterly,* XX (1958), 462–475.

Stevenson, William G. *Thirteen Months in the Rebel Army.* New York: Barnes, 1959.

Still, William N., Jr. "Facilities for the Construction of War Vessels in the Confederacy." *The Journal of Southern History,* XXXI (Aug. 1965), 285–304.

———. "Selma and the Confederate States Navy." *The Alabama Review,* XV (Jan. 1962), 19–37.

———. *The Centennial of Incorporation.* Charleston: The News and Courier Book Press, 1884.

Thomas, Emory M. *The Confederacy as a Revolutionary Experience.* Englewood Cliffs, N.J.: Prentice-Hall, Inc., 1971.

Thompson, Ed Porter. *History of the Orphan Brigade.* Louisville: Lewis N. Thompson, 1898.

Vandiver, Frank E. *Ploughshares into Swords: Josiah Gorgas and Confederate Ordnance.* Austin: University of Texas Press, 1952.

Weller, Jac. "The Field Artillery of the Civil War." *The Military Collector and Historian,* V (Sept. 1953), 29–34.

———. "What We Are Gaining By the War." *Debow's Review, VIII* (April 1862), 327–332.

Wiley, Bell I. *The Life of Johnny Reb: The Common Soldier of the Confederacy.* New York: The Bobbs-Merrill Company, Inc., 1943.

———, ed. *Recollections of a Confederate Staff Officer by G. Moxley Sorrel.* Jackson, Tenn.: McCowat Mercer Press, Inc., 1958.

Winters, Jonn D. *The Civil War in Louisiana.* Baton Rouge: Louisiana State University Press, 1963.

Wise, Jennings C. *The Long Arm of Lee.* New York: Oxford University Press, 1959.